Archaeological Stones and Their Macroscopic Identification

Archaeological Stones and Their Macroscopic Identification

James A. Harrell

Archaeopress Archaeology

ARCHAEOPRESS PUBLISHING LTD
13–14 Market Square
Bicester
Oxfordshire OX26 6AD
United Kingdom
www.archaeopress.com

ISBN 978-1-80327-913-8
ISBN 978-1-80327-914-5 (e-Pdf)

This book is available direct from Archaeopress or from our website www.archaeopress.com

Dedicated to the memory of my grandparents

Fannie Kate (née Dezern) and Jesse Henry Harrell

and

Ethel Bethany (née Smith) and Thomas Carter Sharp

Contents

Part I Minerals

Part II Rocks

List of Figures

List of Tables

Preface

During the over five decades that I have worked with archaeologists, many have expressed an interest in knowing more about how to identify the rocks and minerals they encounter in the field. They recognized that this capability is an important part of their overall understanding of an archaeological site. It is for these people that this book was written. It is hoped that other archaeologists will take a greater interest in rocks and minerals now that the information needed for their recognition has been compiled in one easy-to-use source. This book is the geological reference that I wish had been available to me when I was working on archaeological excavations and surveys in Egypt and other Middle Eastern countries. At the time, my only option was to carry a small library of mineralogy and petrology books with me into the field. Although this book's focus is on archaeological stones, it will also be useful to geology students as a first step in their mineralogy and petrology studies as well as to practicing geologists whose macroscopic identification skills need burnishing.

Two geologists read an earlier draft of this book and offered many useful comments: Dr. V. Max Brown (retired from the University of Toledo, Toledo, USA) and Dr. Leigh Bettenay (at the University of Western Australia, Perth, Australia). It is with gratitude that I acknowledge their assistance as well as that of others who offered helpful suggestions on specific aspects of this book.

Chapter 1: Introduction

1.1 Archaeological Stones

Pottery and worked stone are the most commonly encountered materials in archaeological excavations and surveys. The stones are comprised of 'minerals' and 'rocks', with the latter being granular aggregates of one or more types of minerals (Figure 1.1). The minerals were used as 'gemstones' for jewelry, amulets, seals, and other small decorative objects; sources of metals; and a miscellany of other purposes. Although they occasionally served as gemstones, the rocks were employed mainly as 'building stones' for temples and shrines, palaces and villas, fortresses, administrative and commercial buildings, tombs, roadways, and other monumental structures; 'ornamental stones' for statues and other sculptures, sarcophagi, stelae, funerary vessels, architectural elements in buildings, and other small to large decorative objects; and 'utilitarian stones' for tools, weapons, grinding stones, paint pigments, cooking and serving vessels, loom and net weights, and a large variety of other mundane applications. The identification of these rocks and minerals is an important part of any archaeological study. This goes beyond proper labeling. Knowing a stone's geological identity is necessary for determining its geographical source and, hence, also the trade routes and commercial relationships that brought it to its final resting place.

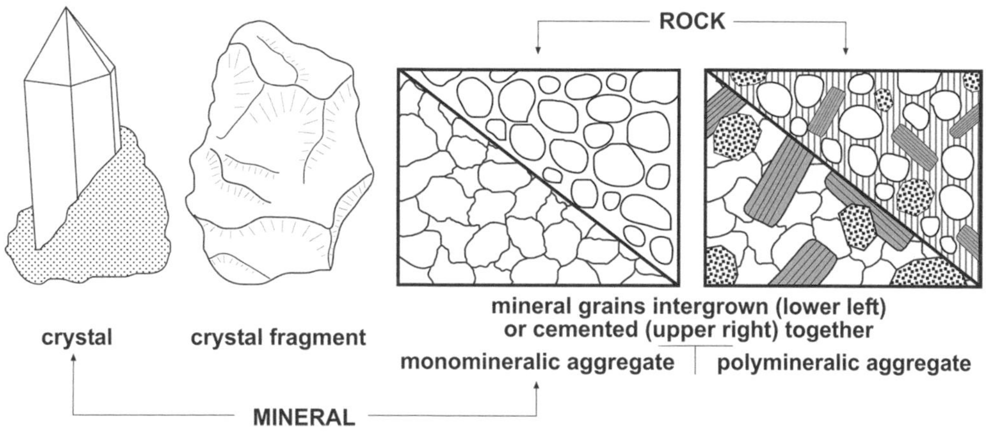

Figure 1.1: The distinction between mineral and rock.

Methodologies requiring sophisticated analytical equipment can be employed to accurately identify the rocks and minerals from archaeological sites, but these are expensive to use and usually require transporting samples long distances, and sometimes across national borders where export issues come into play. While some important samples may get such treatment, the vast majority of stones, for both economical and practical reasons, are identified by macroscopic means in the field or laboratory by members of an archaeological team (Figure 1.2). Sometimes these are geologists, but usually they are archaeologists with little or no formal training in 'mineralogy' (the study of minerals) or 'petrology' (the study of rocks). They have to rely on the many scholarly textbooks and popular guidebooks that have been published on rock and mineral classification and identification. The problem with these references, however, is that they are either too technical in their presentation, depend too much on properties that are difficult to determine, omit some of the useful and easy-to-perform tests, or do not include all the stones and other materials that archaeologists can potentially encounter. The intent of the present work is to avoid these shortcomings while providing archaeologists with a simplified and comprehensive explanation of how to recognize archaeologically important rocks and minerals. Also briefly included in this treatment are other materials found at archaeological sites that are sometimes thought of as 'stones', especially when occurring as small nondescript fragments in soils and sediments, but are technically neither rock nor mineral: e.g pieces of amber, bitumen, bone, ceramics, glass, ivory, and metal. Any of these materials can occur mixed in with the rocks and minerals, and so they need to be identified along with everything else.

Figure 1.2: Identifying a stone at an archaeological site.

The term 'macroscopic' (or, synonymously, 'megascopic') in the context of this book refers to the properties of rock and minerals, and their test responses, that can be observed with the unaided eye or, at most, with a low-power magnifying glass. Such observations are made possible through the use of a relatively small number of inexpensive, simple-to-use tools and tests. These will be explained along with other information needed to identify minerals (Chapter 3) and rocks (Chapter 5). Also provided are determinative keys that guide the identification process. Recognizing rock and mineral varieties is made easier if you understand, in general terms, how they form, and so brief overviews are provided on the origins of minerals (Chapter 2) and rocks (Chapter 4).

Some of the determinative tests presented here are destructive in that they require either a small area on a specimen to be abraded or treated with a drop of dilute acid, or the removal of a tiny fragment for testing in a candle flame, or, after powdering, heating in a test tube. Other tests do no damage and merely require observation (e.g. color, external form, or internal structure) or wetting with water (e.g. to improve visual clarity or to allow measurement of specific gravity). Some specimens, owing to their delicate condition or great artistic or historical value, cannot be subjected to wetting or destructive testing; however, most of these will have a broken surface where some useful observations can be made. Knowing the identity of a rock or mineral is important, and it will often be the case that the minimal damage done to one small area on a specimen is an acceptable price to pay for the knowledge gained.

You may be aware of the several digital applications ('apps') for smart-phones and tablets, where you take a picture of a stone and the app tells you what it is. Some are free, but barrage you with advertisements, and others, with more features and no advertisements, require paid subscriptions. The big question, of course, is: do these really work? If they do, then there would be little need for a book like this one. The answer is: it depends. All these apps do is compare the picture you provide against a large database of pictures of rocks and minerals of supposedly known identity, and then try to find the closest match. This match is the identification of your unknown specimen. Tests have shown the identifications are accurate if you provide nearly textbook perfect pictures of common rock and mineral varieties that exhibit well-developed defining features. They are entirely unreliable, however, for everything else, including unusual varieties and less than ideal examples of the common varieties. This is due to both the incompleteness and inaccuracies of the database, which is made up of labeled pictures scoured indiscriminately from the Internet. Some of these stones are correctly identified but many are not. Additionally, for the pictured stones in the database, some have been correctly labeled, but according to different and often incompatible classification schemes which are not revealed to users and are probably unknown to the app developers. It is also unclear if the backgrounds in the photographs have been blanked out or if they contribute to how the app sees the stones. No doubt, the addition of an 'artificial intelligence' capability (i.e. a machine-learning neural network) to an app will improve its identification accuracy, but it will still be dependent on the quality of the database used to train it. It will, thus, require a considerable leap of faith to trust an app to identify your archaeological stones. It is far better to actually understand what you are looking at and this book will instruct you on how to do this.

Many guidebooks on rocks and minerals – such as the ever-popular ones by Mottana et al. (1978) and Pellant (1992) – include beautiful color photographs of every stone described. While these would seem to be a useful addition, I think they can do more harm than good because the photographs can be both misleading and a disincentive to making close observations. First, it is unlikely that your rock or mineral specimen will look exactly like or, in many cases, even similar to what is shown in the photograph. And second, when photographs are available, people tend to rely overly much on them at the expense of observing the more useful diagnostic macroscopic properties. Color photographs also add significantly to the cost of a book. For all these reasons, photographs are not included in the present work. They are not, in any event, essential for the identification of rocks and minerals.

While the identification process described in this book has been simplified to make it approachable for non-geologists, it cannot be said that it is simple. There is much to learn about rock and mineral properties, and the copious terminology that applies to them. It will also take time to master the observational tests and to be able to deftly navigate the determinative tables. Patience and perseverance, however, will usually guarantee a successful outcome. If you have not already done so, taking an introductory college-level course in 'Physical Geology' will help in providing context and perspective on the potential stone varieties. Alternatively, you can also get some of this background from Allen (2017), an introduction to physical geology written for archaeologists.

1.2 Rock or Mineral?

A basic definition of rocks and minerals has been presented in Figure 1.1, but, in practice, how does one actually recognize whether a given stone is a rock or a mineral? To arrive at this decision, you need to closely examine the specimen (using a hand-lens if necessary) and observe if it is visibly granular: i.e. an aggregate of many grains separated by distinct boundaries. This can be difficult to determine when the specimen's surface is rough, but the fine details can be brought out by wetting the surface with a thin film of water, clear spittle, or, better, mineral oil (the water-soluble laxative).

- If the stone does not appear to be granular, then it is either a mineral or a very fine-grained rock. If it exhibits a crystal form, even a partial one, it is a mineral. It is also a mineral if you can see one or more sets of parallel breakage surfaces throughout the specimen (i.e. cleavage planes as described in section 3.2.3.3). In the absence of these features, it will be difficult to distinguish between a mineral fragment and a rock with grains too small to be seen.
- If the stone is conspicuously granular, then it is a rock. However, if it appears that all (or nearly all) of the grains are the same mineral, then this monomineralic rock can also be considered as the aggregate form of a mineral.

Further observations, as explained in Chapters 3 and 5, will usually reveal the specimen's true identity: i.e. the specific rock or mineral variety.

Part I
Minerals

Chapter 2: Overview of Minerals

2.1 Origin and Nature of Minerals

All the solid, liquid, and gaseous materials making up the Earth are composed of 'atoms', and all atoms that have the same number of positively charged 'protons' in their nucleus belong to the same 'element'. Elements are, thus, defined by their 'atomic number' (i.e. the number of protons). The atoms of a given element can differ in the number of electrically neutral neutrons in their nucleus and this gives rise to different 'isotopes' of an element. The atomic nuclei are surrounded by clouds of negatively charged 'electrons'. There are about 90 elements that occur naturally in the Earth's crust, but only 78 of these are significant (Table 2.1). Two of them, oxygen (O) and silicon (Si), make up 74.3% of the crust by weight. The addition of aluminum (Al) brings the total weight to 82.4%, and another five elements – iron (Fe), calcium (Ca), sodium (Na), potassium (K), and magnesium (Mg) – increases it to 98.6%. Thus, the other 70 elements collectively make up only 1.4% of the Earth's crust. The remaining 12 naturally occurring elements are either short-lived, radioactive varieties or chemically inert noble gases. Elements can lose or gain electrons to become electrically charged 'ions', with the result that positively and negatively charged ions combine to form solid 'crystalline' materials known as 'minerals'.

A mineral exhibits three defining properties: [1] it is solid; [2] it has a crystalline structure (i.e. an orderly and constant three-dimensional arrangement of its constituent atoms); and [3] it has an elemental composition that is either fixed, or, when part of a solid-solution series, varies within defined limits (i.e. the types and relative proportions of its elements are constant enough to be represented by a chemical formula). Atoms of other elements may be present as well, but these are merely occasional impurities and so do not appear in a formula. For many minerals there is more than one way to write their formula while expressing the same elemental composition. For the sake of consistency, the formulae provided in this book are those from the authoritative reference of Klein and Dutrow (2007), and for those minerals not included in this source, the formulae come from *mindat.org*, a highly respected online mineralogy database maintained by the not-for-profit Hudson Institute of Mineralogy. There is another class of natural, solid earth materials that have an amorphous (or at least partially amorphous) internal structure and so technically cannot be considered as minerals. These are the so-called 'mineraloids' and they have only approximate elemental compositions.

Minerals form either by [1] direct crystallization (i.e. precipitation) from a liquid or gaseous vapor; [2] recrystallization (i.e. solid-state replacement) of a pre-existing mineral; or [3] crystallization through biological processes. These processes occur in connection with the formation of the three basic rock types (Figure 2.1): direct crystallization from magma or lava (i.e. molten rock), or hydrothermal solutions in igneous rocks; recrystallization at high (sub-melting) temperatures and/or pressures, or in the presence of chemically active fluids in metamorphic rocks; and both direct crystallization and recrystallization at low temperatures and pressures in the presence of chemically active ground and surface waters in sediments and sedimentary rocks, plus biological crystallization in sediments. For more information on the origins of these rock types, see Chapter 4.

According to *mindat.org*, there are currently about 5800 mineral species recognized (including mineraloids), but only a few dozen of these make up the overwhelming bulk of the Earth's crust and these are the so-called 'rock-forming minerals'. Not surprisingly, the rock-forming minerals consist largely of the eight most abundant elements, especially oxygen and silicon

Table 2.1: Elemental abundances for the Earth's crust.

Atomic Number[1]	Element[2]	Percentage in the Earth's Crust by Weight[3]		Atomic Number[1]	Element[2]	Percentage in the Earth's Crust by Weight[3]
8	oxygen (O)	46.6000		64	gadolinium (Gd)	0.0005
14	silicon (Si)	27.7200		4	beryllium (Be)	0.0003
13	aluminum (Al)	8.1300		35	bromine (Br)	0.0003
26	iron (Fe)	5.0000		55	cesium (Cs)	0.0003
20	calcium (Ca)	3.6300		66	dysprosium (Dy)	0.0003
11	sodium (Na)	2.8300		68	erbium (Er)	0.0003
19	potassium (K)	2.5900		70	ytterbium (Yb)	0.0003
12	magnesium (Mg)	2.0900		72	hafnium (Hf)	0.0003
22	titanium (Ti)	0.4400		32	germanium (Ge)	0.0002
1	hydrogen (H)	0.1400		33	arsenic (As)	0.0002
15	phosphorus (P)	0.1050		42	molybdenum (Mo)	0.0002
25	manganese (Mn)	0.0950		50	tin (Sn)	0.0002
9	fluorine (F)	0.0625		73	tantalum (Ta)	0.0002
56	barium (Ba)	0.0425		74	wolfram (W)	0.0002
38	strontium (Sr)	0.0375		92	uranium (U)	0.0002
16	sulfur (S)	0.0260		53	iodine (I)	0.0001
6	carbon (C)	0.0200		63	europium (Eu)	0.0001
40	zirconium (Zr)	0.0165		65	terbium (Tb)	0.0001
23	vanadium (V)	0.0135		67	holmium (Ho)	0.0001
17	chlorine (Cl)	0.0130		69	thulium (Tm)	0.0002
24	chromium (Cr)	0.0100		71	lutetium (Lu)	0.0001

[1] Number of protons in an element.

[2] Some elements are excluded from this list, such as the noble gases and the short-lived radioactive elements. Following each element's name is its internationally recognized abbreviation or symbol (in parentheses)

[3] Percentages are from Table 5-2 in L.G. Berry, B. Mason and R.V. Dietrich 1983. *Mineralogy – Concepts, Descriptions and Determinations* (2nd edn). San Francisco: W.H. Freeman and Co.

Atomic Number[1]	Element[2]	Percentage in the Earth's Crust by Weight[3]	Atomic Number[1]	Element[2]	Percentage in the Earth's Crust by Weight[3]
37	rubidium (Rb)	0.0090	81	thallium (Tl)	0.0001
28	nickel (Ni)	0.0075	90	thorium (Th)	0.0001
30	zinc (Zn)	0.0070	34	selenium (Se)	<0.0001
58	cerium (Ce)	0.0060	44	ruthenium (Ru)	<0.0001
29	copper (Cu)	0.0055	45	rhodium (Rh)	<0.0001
39	yttrium (Y)	0.0033	46	palladium (Pd)	<0.0001
57	lanthanum (La)	0.0030	47	silver (Ag)	<0.0001
60	neodymium (Nd)	0.0028	48	cadmium (Cd)	<0.0001
27	cobalt (Co)	0.0025	49	indium (In)	<0.0001
21	scandium (Sc)	0.0022	51	antimony (Sb)	<0.0001
3	lithium (Li)	0.0020	52	tellurium (Te)	<0.0001
7	nitrogen (N)	0.0020	75	rhenium (Re)	<0.0001
41	niobium (Nb)	0.0020	76	osmium (Os)	<0.0001
31	gallium (Ga)	0.0015	77	iridium (Ir)	<0.0001
82	lead (Pb)	0.0013	78	platinum (Pt)	<0.0001
5	boron (B)	0.0010	79	gold (Au)	<0.0001
59	praseodymium (Pr)	0.0008	80	mercury (Hg)	<0.0001
62	samarium (Sm)	0.0006	83	bismuth (Bi)	<0.0001

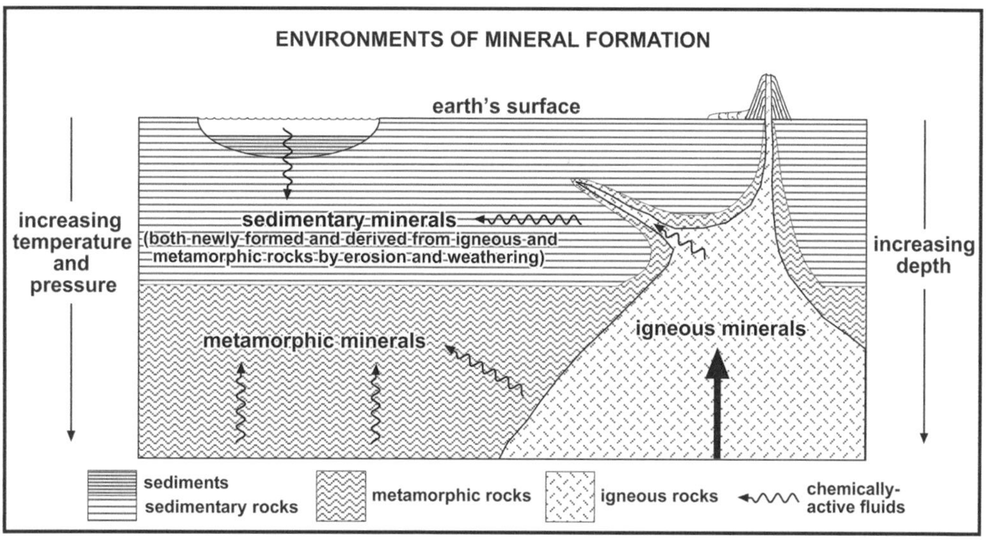

Figure 2.1: Schematic representation of the environments of mineral formation.

2.2 Archaeologically Important Minerals

Each of the standard reference works and textbooks on mineralogy – such as Berry et al. (1983), Gribble (1988), Klein and Dutrow (2007), and Deer et al. (2013) – provides descriptions of a couple of hundred minerals, just those that the authors consider to be the most common or important. The problem with these and other similar books that might be consulted by archaeologists is that many of the minerals they include are not ones of archaeological significance. There is no universally agreed upon list of archaeologically important minerals and no serious attempt has been previously made to formulate one. The list prepared for the present work and appearing in the determinative tables (Tables 3.4-3.6) has been compiled from the principal surveys of archaeological stones, including Hodges (1964), Rosenfeld (1965), Shackley (1977), Ogden (1982), Kempe and Harvey (1983), Herz and Garrison (1998), Rapp (2002), Photos-Jones and Hall (2011), and Siddall (2018). The nearly 150 minerals in the tables also include the common rock-forming minerals because, although they may not be archaeologically significant, they will often be encountered in excavations and surveys.

There are four groups of archaeologically important minerals: [1] those employed as gemstones in jewelry and the decorative arts (i.e. minerals that are at least moderately hard and have an attractive appearance); [2] those used as sources of metal (i.e. minerals that are both common and easily processed to yield metal); [3] those used for paint pigments, and colorants in glasses and vitreous glazes (i.e. minerals that are usually common and produce a colorful powder); and [4] other minerals that were employed for a wide array of mundane applications, including the manufacture of glass, textiles, and leather; medicine; fluxes in metal smelting and refining; pesticides in agriculture; soap; and many others. A mineral's known use in antiquity, if any, is noted in the mineral descriptions in Tables 3.4-3.6.

2.3 Approaches to Mineral Identification

Minerals can usually be identified by their many easily determined macroscopic properties, but more certain identifications, especially for the small mineral grains within rocks, can be made using optical mineralogy or x-ray diffraction (XRD) analysis. In the former, opaque minerals are viewed under a microscope in reflected light using a highly polished slice, and non-opaque (i.e. transparent or translucent) minerals are viewed in transmitted, polarized light, either as finely crushed grains in an immersion oil or as a 0.030 mm-thick slice (i.e. a thin-section). When applied to rocks, this approach is referred to as 'thin-section petrography'. Both the opaque and non-opaque minerals have diagnostic optical properties that can be used to identify them. In XRD analysis, the mineral is ground to a fine powder and irradiated with an x-ray beam. Computer software, which is part of the x-ray diffractometer's instrumentation, analyzes the diffraction pattern produced when the x-rays pass through a rotating sample and suggests an identification based on the information stored in its database. This only works, however, for fully crystalline (non-amorphous) materials. Prior to the latter half of the 20th century, when XRD became commonplace, mineral identification was routinely done by a combination of wet chemical methods and blowpipe analysis (i.e. heating in an intensified flame). This approach required numerous chemicals, many of which were toxic or corrosive. Some of these techniques, however, are safe and easy to do, and so are suitable for the macroscopic identification of minerals described in this book (see section 3.2.11).

Infrared spectrometry (IS) is more limited than XRD in its ability to identify minerals, but has the advantage that it can also recognize amorphous and disordered crystalline materials, such as glass and mineraloids. It requires powdering a sample and exposing it to infrared radiation. The materials' absorbance of this radiation at specific wavelengths depends on the amounts and kinds of chemical bonds, which, in turn, is an indication of the elemental composition and, hence, also the mineralogy. Raman spectrometry (RS) is a non-destructive alternative to XRD and IS, where a laser beam stimulates atomic and molecular vibrations within a mineral, which then scatter the laser light in a diagnostic manner. For archaeologists who lack the requisite petrographical skills and are inexperienced at recognizing macroscopic properties, XRD, IS and RS analyses are the best means of mineral identification. However, these instrumental methods are costly, not always available, and require equipment operators with specialized expertise.

The elemental composition of a mineral specimen is sometimes of interest, either as an aid to identifying it or, more often, as part of a provenance study. In the latter case, the working hypothesis is that impurities (often referred to as 'trace elements' because of their low abundance) within a mineral reflect its geographic source area. A number of quantitative analytical techniques are employed for elemental analyses, including x-ray fluorescence (XRF) spectrometry, combined inductively-coupled plasma–mass spectrometry (ICP-MS), instrumental neutron activation analysis (INAA), atomic absorption spectrometry (AAS), particle-induced x-ray emission (PIXIE) analysis, combined energy- or wavelength-dispersive x-ray spectrometry–scanning electron microscopy (EDS-SEM or WDS-SEM), and electron microprobe (EMP) microscopy among others. Of these techniques, ICP-MS is the most useful overall because of the broad spectrum of elements detected, high precision, and low detection limits. Mass Spectrometry on its own is used to identify elemental isotopes and the ratio of two isotopes in a mineral is often a useful provenance indicator. All these techniques are destructive if used to perform a fully quantitative analysis because they require a sample to be powdered or dissolved. If, however, laser ablation (LA) is used, as is commonly the case for ICP-MS (i.e. LA-ICP-MS), the analysis is essentially non-destructive because the laser vaporizes a miniscule amount of the sample, leaving only microscopic pits. There are also portable x-ray fluorescence (P-XRF) analyzers that are non-destructive and, because they are handheld, are suitable for fieldwork, but they can only provide qualitative or, at best, semi-quantitative data, and just for certain elements.

The vast majority of minerals encountered in archaeological work will need to be identified by macroscopic means using their easily observable physical and chemical properties. These are the subject of Chapter 3. Even with this information and the determinative tables provided, mineral identification is not always easy. The process is mostly a matter of science, but it is also partly an art, where experience and common sense are important. The approach to mineral identification taken in the present work is to key out a mineral using its properties in conjunction with the determinative tables.

Chapter 3: Mineral Properties and Identification

3.1 Equipment and Supplies Needed

An experienced geologist, especially one well trained in mineralogy, can identify the more common minerals at a glance or even from a good photograph. For most people, however, simple tests and close observations are necessary, and for these some inexpensive equipment and supplies are required as listed below.

1. For close-up observation of mineral properties.

   a. Hand-lens (a.k.a. pocket-magnifier or jeweler's loupe). This should ideally have two magnifying lenses, one low power and the other intermediate power, which, when brought together, create a high-power lens (e.g. a hand-lens with 3X and 6X lenses that combine to make 9X). The magnification marked on a lens refers to its 'areal' magnification. The lens' 'linear' magnification is equal to the square root of the areal magnification. For example, a lens labeled as 9X yields a linear magnification of 3 times. For magnifiers with a single lens, most mineralogy references recommend either 5X or 10X power. Wider lenses are better not only because they provide a larger field of view but, more importantly, because they let in more light. A diameter of about 2 cm is ideal and you want lenses that are of sufficiently high quality that they are distortion-free (i.e. provide a 'flat field') all the way to their edge. Unless you are unusually dexterous, a lens that clips to a pair of glasses or can be otherwise worn on the head is best. Plastic lenses are lighter in weight and less expensive than those of glass, but they are more susceptible to scratching and deformation in hot sunlight. If a low- to intermediate-power binocular microscope is available, then this would obviously be better for laboratory work, but a hand-lens is something that you can carry with you at an excavation or during a survey.
   b. Small flashlight with a bright, white LED (light-emitting diode) light.
   c. Mineral oil (in place of water or spittle) for clarifying surfaces.

2. For measuring specific gravity (see also section 3.2.4.2).

   a. Electronic balance. A small, battery-powered model is sufficient provided it can reproducibly weigh a load to the nearest hundredth of a gram. Such an instrument will almost always be available where archaeological work is being done and is usually to be found with the finds registrar. Alternatively, a high-quality spring scale may be used for less precise weights.
   b. Cup or other small container for water.
   c. Nylon thread, as thin as possible and preferably colored.
   d. Liquid household detergent.

3. For measuring Mohs scratch hardness (see also section 3.2.3.1).

   a. Small, steel pocket knife (a.k.a. penknife) or other steel tool.

b. Small square of normal window/plate glass.
c. Small square of unglazed, white porcelain. If not purpose-made for testing minerals, the unglazed back of a ceramic floor or wall tile will also serve.
d. Small piece of metallic copper (Mohs 3).
e. Crystals of apatite (Mohs 5), orthoclase (Mohs 6), and quartz (Mohs 7) for determining the hardness of the steel tool, glass plate, and porcelain tile in a-c.
f. In place of items a-e, a set of Mohs hardness index minerals or a set of styluses' with metal tips having the same hardness as the index minerals.

4. For powdering a mineral either to observe its streak color or its reaction to acid or heat (see also sections 3.2.2.2, 3.2.8, 3.2.10, and 3.2.11.4).

a. Small steel hammer and small slab of hard steel to use as an anvil.
b. Silicon carbide (carborundum), tungsten carbide, or diamond scriber.
c. Small square of unglazed, white porcelain as in 3c above.

5. For testing carbonate minerals (see also sections 3.2.8 and 3.2.9).

a. Dilute (5% or 10%) hydrochloric acid in a small glass bottle with a dropper cap or a plastic squeeze bottle with a sniffer top.
b. Alizarin Red S dye for staining.

6. For testing fusibility and other heat-dependent properties (see also section 3.2.11)

a. Candle or any wicked lamp burning olive oil, alcohol, or other liquid fuel, or a butane or propane gas burner, including a cigarette lighter. Candles are the least expensive and most readily available option. They do, however, have one disadvantage: they leave a black, oily soot on whatever is heated in the flame, but this is easily wiped off and does not occur if a blowpipe is used. At least a few candles are needed (and do not forget the matches).
b. Blowpipe for producing an intensified flame.
c. Steel forceps or tweezer for holding specimens in a flame.
d. Pyrex test tubes (any size will do) and Pyrex 20 ml beakers for heating and chemical tests. At least a few of each of these easily broken items will be needed.
e. Wire-clamp or other holder for test tubes and small beakers.
f. Cylindrical brush for cleaning test tubes.

7. For testing magnetism (see also section 3.2.5): small, strong magnet of the rare-earth type (e.g. neodymium). These come in a wide variety of shapes and sizes, but a cylindrical, rod-like magnet ~1 cm in length is ideal.

8. For testing isotropy (see also section 3.2.2.4)

a. Two polarized filters, either plastic sheets or glass plates. The two lenses from authentic Polaroid sunglasses will also work.
b. Aluminum foil or a short, cylindrical tube of any opaque material.

9. For testing frictional electricity (see also section 3.2.6): a piece of wool or silk cloth.

10. For testing fluorescence and phosphorescence (see also section 3.2.2.5): short- and long-wavelength ultraviolet ('black') lights (254 nm and 365 nm, respectively). These are available as small, battery-powered lights, including some that are no larger than a ballpoint pen.

3.2 Macroscopic Properties and Tests

3.2.1 Habit

The physical form a mineral takes is called its 'habit' or 'morphology'. When minerals grow into a fluid- or vapor-filled space, such as within a magma body or a cavity inside a rock, they will usually develop into distinctive forms called 'crystals' (see section 3.2.1.1). Crystals can sometimes also develop within solid rock during the recrystallization of pre-existing mineral grains. Crystal form is an external expression of a mineral's internal atomic structure. Well-developed crystals are uncommon, however. Many minerals occur in crystalline aggregates where the crystal form is not expressed (see section 3.2.1.2). Crystallization order is partially to blame for this, with the earlier-precipitated grains showing better crystal form than the later ones that are unable to grow unhindered. This phenomenon is especially pronounced when two or more minerals crystalize sequentially as the conditions change (e.g. with declining temperature in magma, feldspar precipitates before quartz and, therefore, shows better crystal form in the resulting rock).

3.2.1.1 Crystals

Each mineral species belongs to one of six 'crystal systems', where each system is characterized by a set of three-dimensional axes that differs from other systems in their relative lengths and angular relationships or, in one case, the number of axes. It is each system's unique symmetry that determines the general forms that its crystals can take and these are a useful diagnostic property for mineral identification. With one exception, crystals are characterized by a generally symmetrical arrangement of a number of flat 'faces' that represent internal planes of high atomic density (i.e. the planes with the highest density of atoms in a crystal's structural lattice). The exception is the hair-like filiform (a.k.a. capillary or wiry) crystals that are the product of unusually rapid growth along one axis combined with rotations of successive structural lattices known as 'screw dislocations'. The six crystal systems are defined as follows (see Figure 3.1 for graphical representations of the sets of axes, and Figure 3.2 for illustrations of the common generalized crystal forms).

Isometric (or **Cubic**): the three axis (a_1, a_2, and a_3) are of equal length and are all mutually perpendicular (i.e. intersect at 90° angles). The crystals usually have either a square or polygonal circular outline when viewed parallel to any axis, but the outline can also sometimes be triangular. The most common general crystal forms are cubical, tetrahedral, octahedral, dodecahedral, pyritohedral, tetrahexahedral, and trapezohedral.

Tetragonal: two of the axes (a_1 and a_2) are of equal length and the third (c) is either shorter or longer than the others, and all axes are mutually perpendicular. The crystals usually have a square or rectangular outline when viewed parallel to any axis, but octagonal or diamond-shaped outlines can also occur. The most common general crystal forms are prismatic, tabular, and dipyramidal. Twinned crystals (see below) can exhibit a pseudo-tetrahedral form.

Orthorhombic: the three axes (a, b, and c) are all of unequal lengths and all are mutually perpendicular. The crystals usually have either a rectangular or oval- to diamond-shaped outline when viewed parallel to any axis. The most common general crystal forms are prismatic, tabular, and dipyramidal. Twinned crystals can exhibit pseudo-octahedral, and pseudo-hexagonal prismatic and tabular forms.

Hexagonal (includes the **Trigonal** sub-system of some mineralogists): there are four axes, three (a_1, a_2, and a_3) are of equal length, lie within a single plane, and are separated by 60° angles (which produces 120° angles between adjoining crystal faces, as in Figure 3.3). The fourth axis (c) is either shorter or longer than the others and is perpendicular to them. The crystals usually have a hexagonal or, less often, a triangular outline when viewed parallel to the c axis. The most common general crystal forms are prismatic, tabular, dipyramidal, scalenohedral, and rhombohedral. Twinned crystals can exhibit a pseudo-cubical form.

Monoclinic: the three axes (a, b, and c) are of unequal lengths with two of them (a and c) intersecting at an oblique angle and the other axial pairs (a-b and b-c) mutually perpendicular. The crystals resemble those of the orthorhombic system except that the matching (parallel) faces on opposite sides of a crystal are not perpendicular to the adjoining faces. The most common general crystal forms are prismatic (mostly short or stubby) and tabular. Twinned crystals can exhibit pseudo-cubical, pseudo-octahedral, pseudo-hexagonal prismatic and tabular, and pseudo-rhombohedral forms.

Triclinic: the three axes (a, b, and c) are of unequal lengths and intersect each other at oblique angles. The crystals are highly variable in outline, and their most common general forms are prismatic, tabular (usually square or rectangular in outline), and dipyramidal. Twinned crystals can exhibit pseudo-cubical, pseudo-hexagonal tabular, and pseudo-rhombohedral forms.

In the real world, crystals are rarely perfectly symmetrical forms with uniformly developed faces. They usually exhibit an unequal development of their primary crystal faces due to different growth rates for each, and they can also possess secondary crystal faces that cut across the primary ones (Figure 3.3). The angular relationships between the primary crystal faces remain unchanged, however. Despite the great variety and complexity of crystals, the general forms shown in Figure 3.2 are usually discernable within the more complex natural forms. When one mineral species replaces a different, pre-existing mineral species in a rock, it sometimes happens that the original crystal form is preserved. Such secondary growths are referred to as 'pseudomorphs': they look like one kind of mineral but, compositionally, are actually another mineral (e.g. goethite replacing cubical pyrite or dodecahedral almandine garnet).

Two or more symmetrically positioned crystals of the same mineral are said to be 'twins' if they share atoms across a plane or a volume of space. Twins are of three basic types: (1) penetration – two crystals appear to pass through each other (i.e. they share a volume of space); (2) contact – two crystals are attached to each other along a plane; and (3) repeated

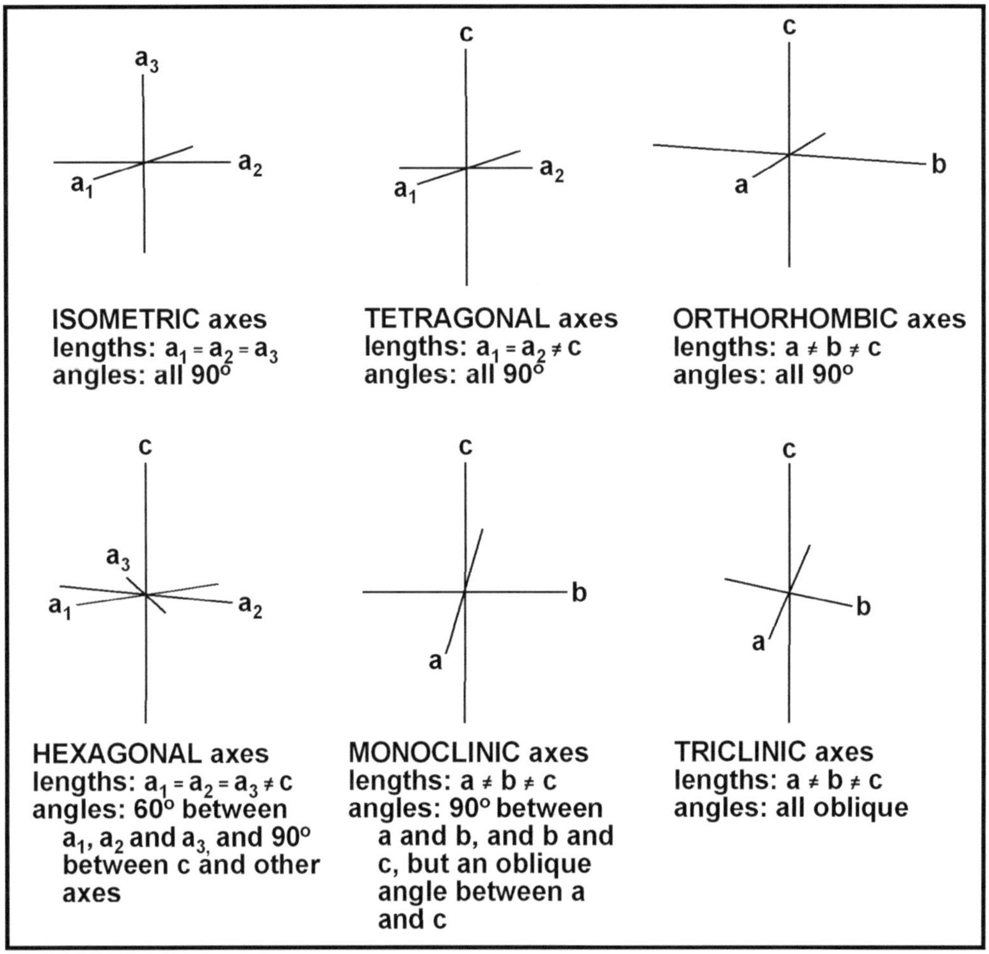

Figure 3.1: The six mineral crystal systems.

– three or more crystals are attached in a parallel series of contact twins. Twinned crystals can take many varied, and at times bizarre, forms that are mineral specific. Figure 3.4 shows examples of twinning for some of the common minerals.

The crystals of some minerals occur in distinctive groupings and several of these are recognized (Figure 3.5).

Drusy: a surface covered by closely spaced, small crystals.

Geode: a hollow or partly hollow rock with the inner surface covered by drusy crystals.

Radiated (or **Divergent**): long prismatic, usually acicular or bladed, crystals that radiate outward from a central point. When radiated acicular crystals have a feathery or plume-like appearance, they are described as **plumose**.

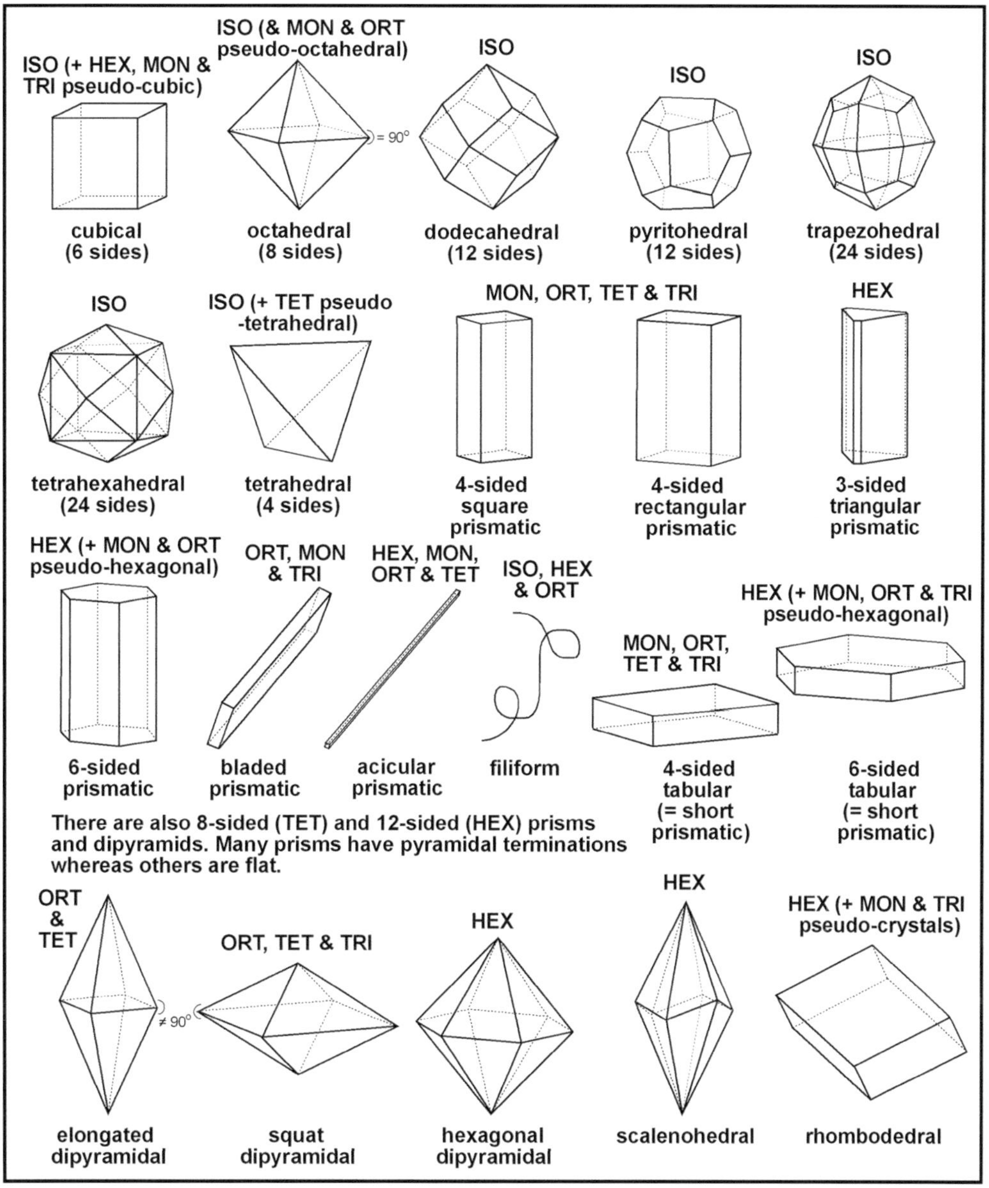

Figure 3.2: The most common generalized forms taken by crystals in the six crystal systems (ISO = isometric, TET = tetragonal, ORT = orthorhombic, HEX = hexagonal, MON = monoclinic, and TRI = triclinic).

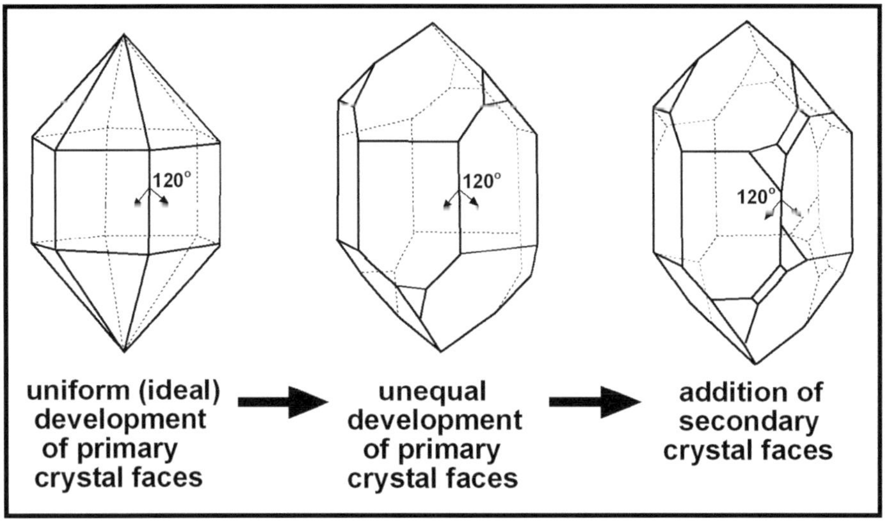

Figure 3.3: Development of crystal faces in quartz, a mineral belonging to the hexagonal system.

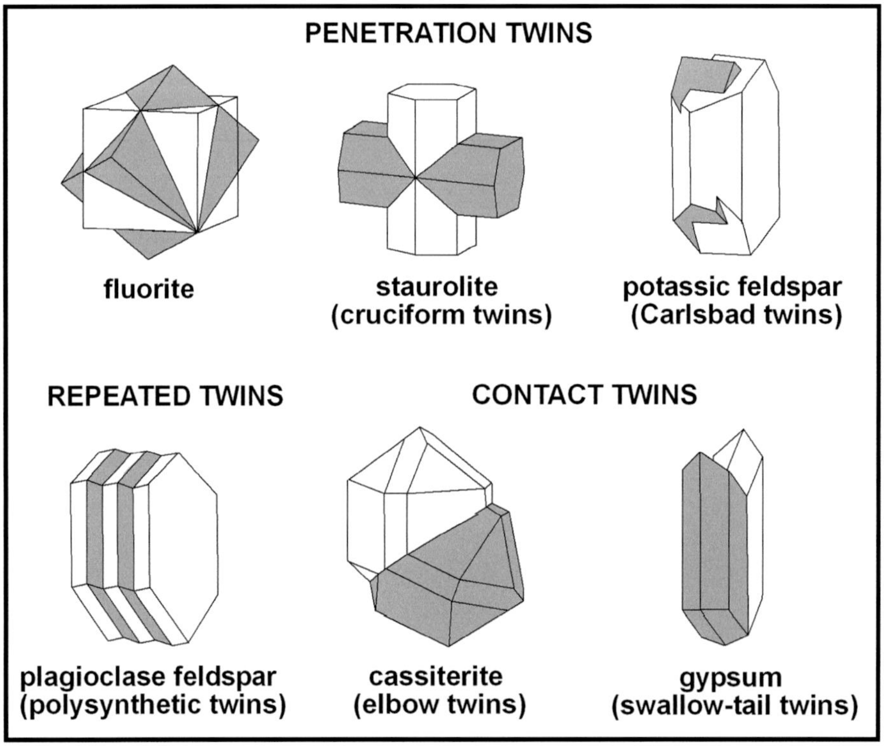

Figure 3.4: Twinning in some common minerals.

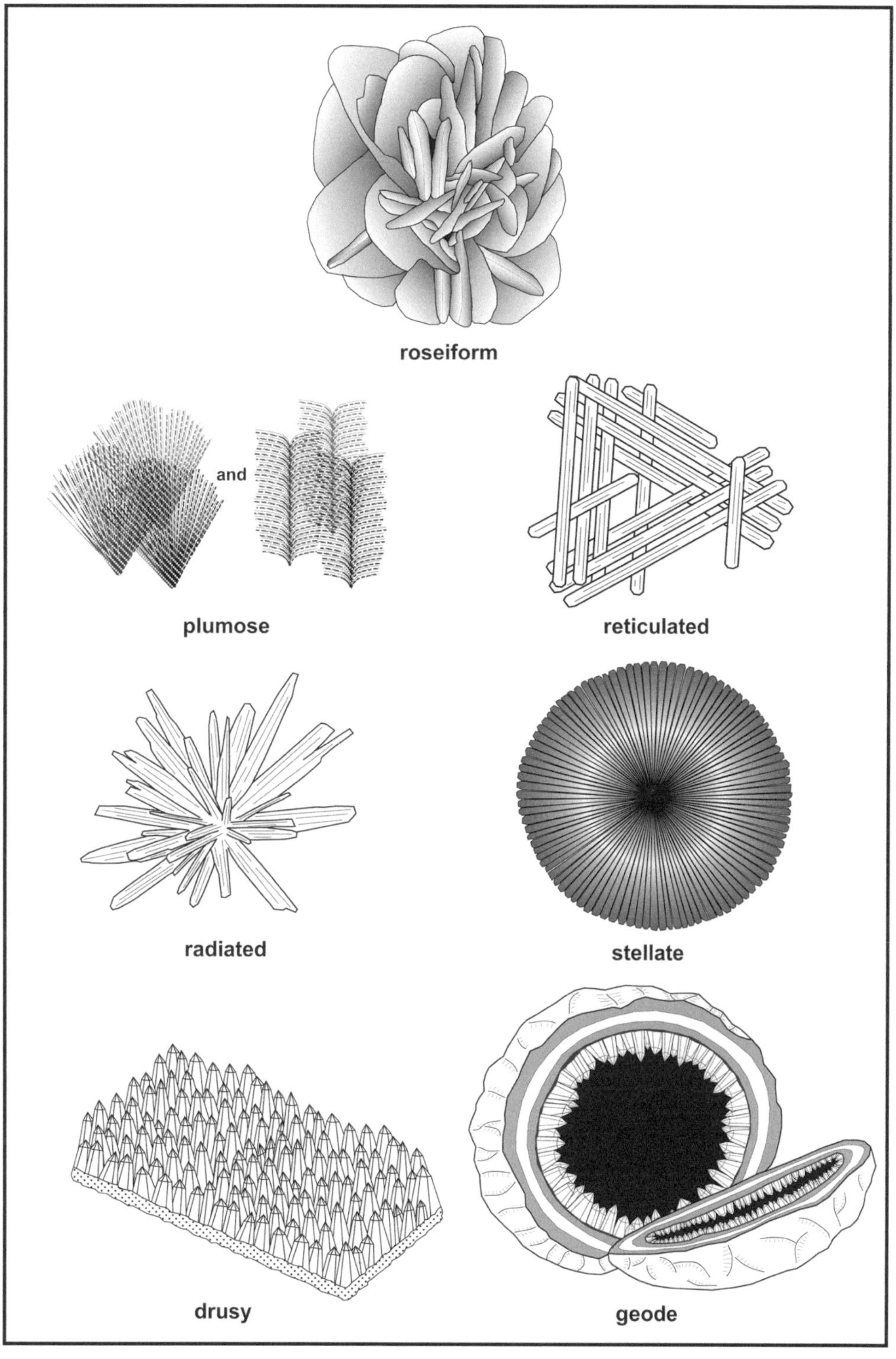

Figure 3.5: Common forms taken by groups of crystals.

Reticulated: long prismatic, usually acicular or bladed, crystals that form a crisscrossing, lattice-like arrangement.

Roseiform (or **Rosette**): a series of interpenetrating bladed or tabular crystals that resemble petals in a flower-like arrangement.

Stellate: a series of long, usually acicular, prismatic crystals that occur in a flat, roughly circular arrangement resembling a starburst.

3.2.1.2 Aggregates

In addition to forming readily visible, solitary or grouped crystals, many minerals also occur as intergrown, microscopic to macroscopic grains without crystal form in an, ideally, monomineralic aggregate. These can exhibit a variety of internal structures and external forms.

Internal Structure (Figure 3.6)

Granular: an aggregate of mineral grains of similar size and roughly equidimensional form. Grain size will vary from one specimen to another and is often described in relative terms like microscopic (or, equivalently, compact), and fine to coarse macroscopic. The term **saccharoidal** is sometimes used to describe a granular texture where the grains are of a size similar to the crystals in granulated sugar.

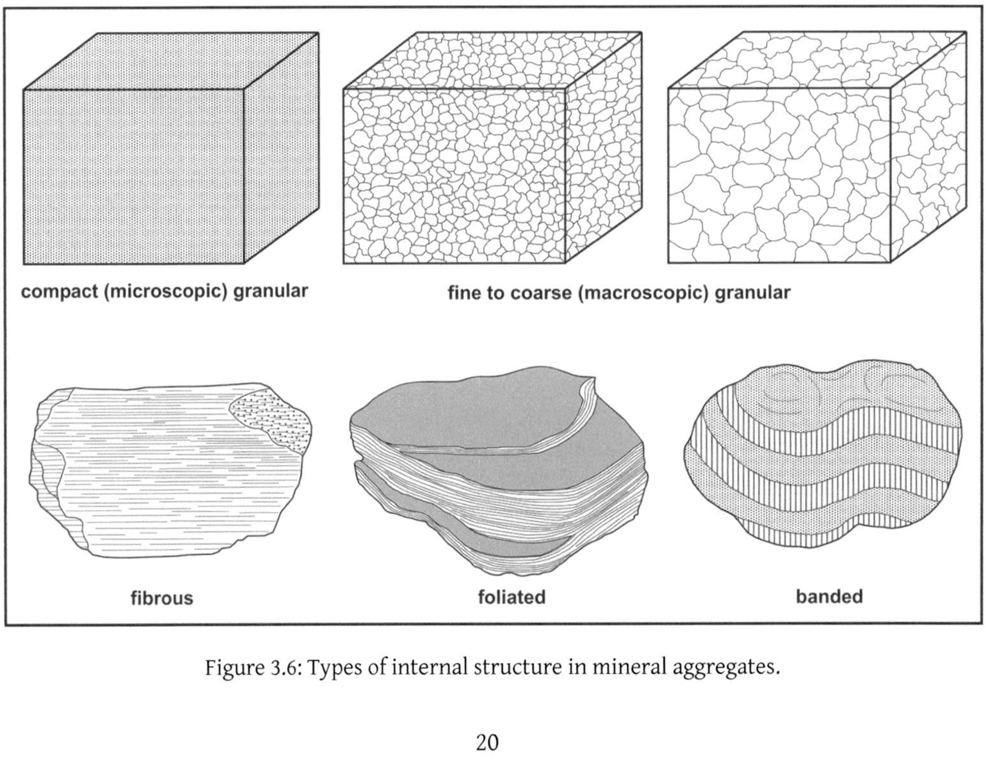

Figure 3.6: Types of internal structure in mineral aggregates.

Fibrous (or **Asbestiform**): an aggregate of closely packed, parallel fibers which can be either microscopic or macroscopic.

Foliated (or **Micaceous**): an aggregate of parallel, platy grains that can usually be separated into thin sheets or scales. The grains can be either microscopic or macroscopic.

Banded (or **Lamellar**): an aggregate exhibiting thin, parallel, alternating layers of different color or texture. The layers can have either a granular or fibrous internal structure. Such an aggregate is also described as **concretionary** when the bands are arranged concentrically around a central point.

EXTERNAL FORM (Figure 3.7)

Colloform: well-rounded, bulbous or globular forms in an internally banded, granular to usually fibrous aggregate with a concentric structure where the fibers are perpendicular to the banding. There are three main sub-varieties.

Botryoidal: a mass resembling a bunch of grapes.

Mammillary: a rounded, breast-like mass.

Reniform: a rounded, kidney-shaped mass.

Stalactitic (or **Columnar**): slender, tapering conical or cylindrical masses resembling cave stalactites. The term **helictitic** is applied when stalactitic forms are curved or twisted. As with the colloform aggregates, these usually have a radially fibrous, concentrically banded internal structure.

Oolitic and **Pisolitic**: numerous sand-size (< ~2 mm; oolitic) or pebble-size (> ~2 mm; pisolitic), well-rounded spherical to ovate forms within a sedimentary rock or soil. As with the colloform aggregates, these usually have a radially fibrous, concentrically banded internal structure.

Dendritic (or **Arborescent**): a branching, tree-like form.

Massive: no distinctive or characteristic form.

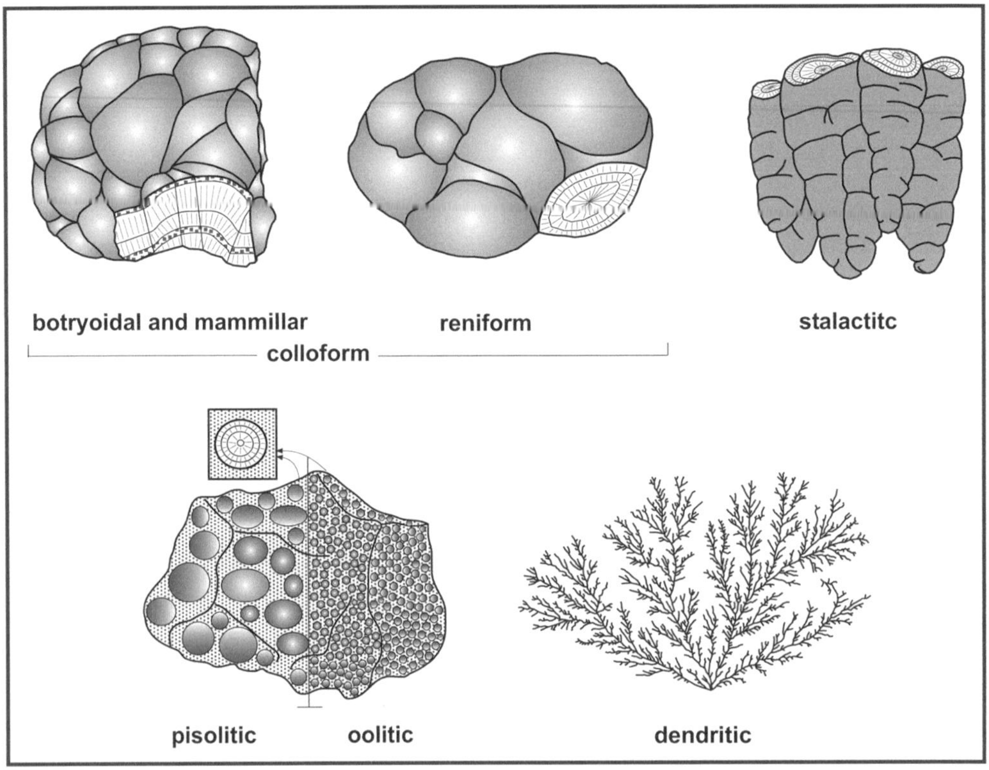

Figure 3.7: Common external forms taken by mineral aggregates.

3.2.2 *Phenomena Dependent Upon Light*

3.2.2.1 Mineral (Whole) Color

A mineral's color is determined by the wavelengths of light it transmits or reflects. Some minerals will show a different color when viewed with light transmitted through them from behind than when viewed with light reflected from an incident source in front. A mineral's color is determined by its elemental composition and internal structure, but can be variable depending on the impurities present. The 'whole' color is a mineral's most obvious property, but it is not always a reliable one for identification. There are two reasons for this. First, many minerals are naturally highly variable in color because of impurities. And, second, color is subjective in that no two people are likely to describe it in exactly the same terms. To avoid this subjectivity, you might expect that mineral colors are reported using standard color charts for comparison (e.g. one employing Munsell color chips), but this has not been done by mineralogists. Mineral color is best observed under a 'white light' source, like the sun, because some kinds of artificial lighting can contribute their own color characteristics to a mineral. Generally speaking, bright white LED lights are adequate for illuminating minerals without altering their colors, but natural sunlight is always best.

When wetted with water, a mineral's color will be a darker shade than when dry. This is due to the higher reflectivity of the wet surface (i.e. less light is absorbed and then reemitted by the mineral). For the same reason, a polished mineral surface appears darker than a rough one. The color of many minerals is unstable and so changes over time. For example, most metallic minerals quickly tarnish when exposed to the atmosphere due to reactions with mainly oxygen but sometimes also with other elements like sulfur. The ultraviolet radiation in sunlight will fade the color of some nonmetallic minerals. The colors of minerals with either water (H_2O) or hydroxyl (OH) in their chemical formula can change due to partial dehydration (i.e. removal of H_2O or OH) caused by heating in sunlight or a fire, or to rehydration caused by immersion in water or humid air. Complete dehydration or hydration often results in the formation of an entirely different mineral (e.g. yellowish limonite, ~$FeO(OH){\bullet}H_2O$, can dehydrate to become reddish hematite, Fe_2O_3). It is only the dry, unaltered mineral colors on unpolished surfaces that are reported in the determinative tables of this book (Table 3.4-3.6), but the tarnished colors of the metallic minerals are also provided when these can be useful for identification.

3.2.2.2 Streak (Powder) Color

If a mineral is firmly dragged across a rough (unglazed) porcelain plate and its hardness is less than that of the plate's (usually 6.5-7 on the Mohs Scale; see section 3.2.3.1), a powdery residue will be left behind. This is the mineral's 'streak'. Care should be taken to ensure that the residue is not coming from the porcelain plate, as would be the case if it is softer than the mineral being tested. Small, white squares of unglazed porcelain are sold as 'streak plates' for mineral testing, but you can also use the rough back of any glazed ceramic tile from a hardware store. Such tiles are always very hard, but to varying degrees, ranging between 5 and 7 on the Mohs scale. Consequently, you should always determine the hardness of a streak plate before using it (see section 3.2.3.1 for further discussion).

If unglazed porcelain is not available, crush the mineral to powder between an anvil and hammer (or two hard stones) and smear it across a white sheet of paper to reveal the 'streak color'. This approach has the advantage of working for all minerals regardless of their hardness. Alternatively, you can use a silicon carbide or a less expensive tungsten carbide scriber (Mohs = 9.5 and 9, respectively) to powder a much smaller portion of a mineral by simply scratching it. The powdery residue so produced exhibits the streak color. Scribers, which are commonly used for inscribing glass and metals, are widely available in hardware and craft stores. Diamond scribers are also available, but are more expensive and have tips that are too easily broken. The same principle as the scriber applies to a steel knife blade when this is used to scratch a softer mineral.

The powdered forms of most minerals exhibit either different shades of the same color or sometimes entirely different colors than the whole minerals. This is because more light is transmitted through the thinner particles in a powder than through larger grains or crystals, which, with their impurities and other imperfections, will have different light absorption and reflection properties. Whereas a mineral's whole color may vary, its powdered color is always the same, making it a useful diagnostic property for mineral identification. As a consequence of the effect of particle size, most transparent and translucent minerals, even if deeply colored when whole, produce a white powder. Opaque minerals with a non-metallic luster usually produce a powder of the same color but with a lighter shade than the whole mineral, and opaque minerals with a metallic luster produce a powder that is either darker or a different color than the whole mineral.

3.2.2.3 Luster

A mineral's luster (a.k.a. 'shine') refers to the amount and quality of light reflected from its surface. It is a property that depends on a mineral's internal structure, surface roughness, and diaphaneity (i.e. whether transparent, translucent, or opaque; see section 3.2.2.4). The recognition of luster is also affected by the observer's perceptions based on their experiences with commonplace materials. Consequently, except for the distinction between metallic and nonmetallic lusters, this is not a particularly useful property for the nonmetallic minerals because it is highly subjective and different specimens of the same mineral can exhibit more than one kind of luster. The different kinds of luster used in mineral descriptions are as follows.

Metallic: metal-like and exhibited by some opaque minerals. A metallic luster is the result of a highly reflective surface (caused by the mineral's high opacity) together with what is normally perceived as a metallic color (usually shades of gray or yellow).

Submetallic: imperfect or dull metallic luster and exhibited by some opaque minerals. This will be seen as metallic by some observers and nonmetallic by others. The distinction is a problem especially for minerals that are opaque and reflective, but have a dark color not normally associated with metals (usually black). It is also an issue for other opaque minerals with a pitted, fine-grained, or otherwise irregular surface that interferes with the reflection of incident light.

Nonmetallic: exhibited by all transparent and translucent minerals. Eight kinds of nonmetallic luster are commonly recognized.

Vitreous: shiny like broken glass. This is by far the most common luster for nonmetallic minerals.

Adamantine: shiny like a cut diamond. This is similar to a vitreous luster but has a harder, more sparkling and brilliant shininess.

Resinous: shiny like tree resin. This is characteristic of vitreous or adamantine minerals with resin-like colors (i.e. shades of brown, orange, yellow, or red).

Pearly (or **Nacreous**): shiny like a pearl. This is typical of minerals with a well-developed internal lamellar structure due to either prominent cleavage in one direction (see section 3.2.3.3) or a foliated habit in a monomineralic aggregate (see section 3.2.1.2). Light enters the mineral and then reflects from multiple lamellae and thereby produces a diffuse, pearl-like glow.

Greasy: shiny like a surface covered with a thin film of oil or grease. This is usually caused by either a well-developed microscopic surface roughness or abundant microscopic inclusions of other minerals.

Waxy: shiny like a surface coated with a thin film of candle wax. This is gradational with the greasy luster and is typically exhibited by the finest-grained (always microscopic) monomineralic aggregates or by amorphous mineraloids.

Silky: shiny like silk cloth. This is typical of minerals that occur as aggregates of fine, parallel fibers.

Dull (or **Earthy**): a non-reflective surface like loose soil. This is exhibited by a monomineralic aggregate that is either finely granular or highly micro-porous, with both conditions causing the surface to scatter light in all directions rather than reflecting most of it toward the observer. A sub-variety of this kind of luster is referred to as **pitchy** by some mineralogists and presents a dull, black surface like that of hardened asphalt (i.e. pitch).

3.2.2.4 Diaphaneity and Isotropy

The ease with which light passes through a mineral is its **diaphaneity** (a.k.a. 'transparency' or 'opacity'). Put another way, diaphaneity is the quantity of light transmitted through a mineral. It is best observed along the thin edges of specimens that are illuminated from behind by a bright light source, either the sun or a LED. There are three main categories and two sub-categories of diaphaneity.

Transparent: the outline of an object is distinct when viewed through the mineral (like looking through a clear glass pane).

Semi- or **Sub-Transparent**: the outline of an object is blurred when viewed through the mineral (like looking through a slightly milky plastic sheet).

Translucent: no object can be seen but light is transmitted through the mineral (like looking through a thick, cloudy glass pane or thick plastic sheet).

Pseudo-Opaque (or **Semi-** or **Sub-Translucent)**: only the thinnest edges of a mineral transmit light.

Opaque: no light is transmitted, even through the thinnest edges (like looking at any metal).

While diaphaneity would seem to be a straightforward property to recognize, it is complicated by the size and shape of a mineral specimen. For example, what appears to be translucent or opaque in a thick section might look transparent or translucent, respectively, in a thin one. This is why it is important to focus only on the thin edges whenever possible. When such edges are lacking, as in well-rounded specimens, you have to look tangentially along a rounded edge. For this to work, however, the light source must be very bright. It is because of these issues that there is some subjectivity in describing diaphaneity. The usefulness of this property is also lessened by the fact that a given nonmetallic mineral will often exhibit a range in diaphaneity among different specimens.

Transparent and translucent minerals will transmit light in either all directions (isotropic) or only in certain directions (anisotropic). A simple test of **isotropy** is to overlay two polarized filters that are rotated relative to each other such that all transmitted light is blocked (i.e. the stacked filters look black). A strong light source behind the filters is required, either sunlight or a bright white LED. A mineral specimen, the thinner the better, is then placed between the 'crossed polarizers'. There must not be any light entering between the polarizers; thus, the test is best done outside at night or in a dark room. Alternatively, if there is strong ambient light, either cover the gap around the polarizers with aluminum foil or place a short, opaque cylindrical tube between them with the specimen inside. If no light is transmitted through the stack of filters and specimen (i.e. the specimen will either be invisible or, at most, show a dimly lit outline), then the mineral is isotropic. If some light makes it through, then the mineral is anisotropic (i.e. it will appear brightly lit). All minerals in the isometric crystal system (see section 3.2.1.1) are isotropic whereas all minerals in the other five crystal systems are anisotropic. Amorphous materials, like glass and mineraloids, are also isotropic.

3.2.2.5 *Other Optical Effects*

In Reflected Light

Some transparent and translucent minerals show variations in color when turned under a strong, reflected-light source with the principal variations as follows.

Asterism: a scattering of light into star-like rays in multiple directions.

Chatoyancy: a silky sheen with a shifting band of light.

Iridescence: a vivid, rapidly changing play-of-colors.

Labradorescence: a slowly changing play-of-colors primarily in different shades of blue.

Opalescence: a shifting hazy, milky or pearly reflection emanating from the interior of a specimen.

In Transmitted Light

When the color of a transparent or semi-transparent mineral changes as it is turned in front of a strong transmitted-light source, it is said to exhibit **pleochroism**. There can be two or, less commonly, three different alternating colors or different shades of one color.

From Induced Light

Luminescence is an induced emission, rather than a passive reflection or transmission of light. The emitted light is sometimes faint and so the tests are best done in a darkened room or outside at night. Luminescence can take one of the following four forms.

Fluorescence: a light emission triggered by invisible ultraviolet (UV) or other high-energy electromagnetic radiation. For ordinary mineral testing, a short- or long-wavelength UV light is used (a.k.a. 'black light' with, respectively, 254 nm and 365 nm peak wavelengths).

This type of luminescence is especially well displayed by the mineral fluorite, which is named after the phenomenon.

Phosphorescence: the same as fluorescence except the light emission continues after removal of the exciting radiation.

See Tables 3.7 and 3.8, respectively, for the fluorescence and phosphorescence colors of luminescent minerals in Tables 3.4-3.6 and 5.18.

Triboluminescence: a light emission triggered by the rapid crushing, scratching, striking, or rubbing of a mineral.

Thermoluminescence (a.k.a. **pyroluminescence**): a light emission triggered by heating a mineral to below its 'red heat'.

Of these phenomena, fluorescence is the most useful for mineral identification. It arises from either the presence of certain elements in a mineral's chemical formula or, more often, from elemental impurities (usually the so-called 'transition metals'). Because there can be different amounts and types of impurities among specimens of a given mineral, its fluorescence can vary in intensity and color. Also, some minerals fluoresce only in short- or long-wavelength UV light, and some fluoresce in both, with either the same or different colors and intensities. Additionally, the intensity of a mineral's fluorescence is usually increased after heating in fire and this is especially true of calcite. The vast majority of minerals are not fluorescent, but care must be taken when testing them. They usually appear black under a UV light, but the UV lamps also emit some visible violet light and its reflection off a mineral, especially a light-colored one, can be mistaken for violet fluorescence. Although fluorescence of this color does occur, it is very rare. If you are uncertain whether you are seeing violet fluorescence or reflection of violet light, these can be distinguished by a simple test for the short- (but not long) wavelength UV light. Place a piece of clear, colorless glass between the specimen and lamp and this will block all the short-wavelength UV light. If the specimen still has a violet glow, then this is reflected light. Because it is so variable and most minerals do not exhibit it, fluorescence is more of a confirmatory test than an exploratory one.

3.2.3 Strength and Breakage

3.2.3.1 Mohs (Scratch) Hardness

The ease with which a mineral can be scratched is a fundamental and constant property and, therefore, is an especially important one for mineral identification. A 'scratchability' scale was first proposed in 1822 by the German mineralogist Friedrich Mohs and is still in use today. This is the 'Mohs Hardness Scale' (Table 3.1) and it employs ten index minerals numbered 1 to 10, with the scratch resistance increasing with the number. The difference in scratch hardness across the scale is not uniform, however. The increases in hardness are small and similar between 1 and 4, and larger and less similar between 4 and 9, with the hardness difference between 9 and 10 as great as that between 1 and 9. Nevertheless, the scale remains a useful and objective means of distinguishing minerals. It is important to keep in mind, though, that the scratchability of a mineral is not the same as its resistance to breaking.

A diamond, for example, has the highest scratch hardness of any natural earth material but it is easily shattered when tapped with a hammer.

Table 3.1: Mohs scratch hardness scale

Mohs Hardness	**Index Minerals**	**Commonplace Materials**	
1	talc	pencil lead (1-1.5)	smudges paper and fingers
2	gypsum	fingernail (2-2.5)	
3	calcite	copper and bronze (3-3.5)	
4	fluorite	common steel nail (4-5)	
5	apatite		
		steel knife blade (usually 5-5.5 but can range up to 6.5) and common plate/window glass (5.5-6)	
6	orthoclase		
		concrete/masonry steel nail and common steel file (~6.5)	
		unglazed porcelain streak plate (6.5-7)	
7	quartz		
8	topaz		
9	corundum	tungsten carbide scriber (9)	
		silicon carbide (carborundum) scriber (9.5)	
10	diamond		

The Mohs hardness of a mineral is determined in the following way. One begins by testing with a soft index mineral and then works sequentially through harder ones, or vice versa from the other end of the scale. Each index mineral is firmly pulled or pushed across the surface of a specimen of unknown hardness. When possible, use a sharp corner or edge on the index mineral. If, for example, the specimen is scratched by quartz (Mohs 7) but not by orthoclase (Mohs 6), then one concludes that its hardness is between 6 and 7. The hardness can often be further refined, however. If the specimen can be scratched by quartz only with difficulty in comparison to the ease with which quartz scratches orthoclase, then its hardness is 6.5-7. If, on the other hand, the quartz easily scratches the specimen but the specimen scratches the orthoclase with difficulty, then it has a hardness of 6-6.5. Two different minerals of the same hardness can just barely scratch each other. If this is the case for quartz and the unknown mineral, then the latter can be said to have a hardness of 7. And finally, if the specimen is easily scratched by quartz and it in turn easily scratches orthoclase, then its hardness is 6.5. Saying that a mineral has a hardness of 6.5 does not mean that it is exactly midway in hardness between 6 and 7, only that it is somewhere between these values and not close to either one. The same determination process applies, of course, to an unknown mineral with a hardness falling between any two other index minerals on the scale. Some mineralogists argue that a hardness value falling between two index minerals should be reported, for example, as 6½ rather than 6.5, because the latter notation implies a measurement precision that does not exist. I think this is being overly fussy and so the decimal notation will be used in this book. Mohs hardness is provided for the minerals in Tables 3.4-3.6, and these minerals are also listed in order of increasing hardness in Table 3.10.

There are several things to consider when applying the Mohs hardness test. First, only scratch surfaces that are fresh (unweathered) and clean, and that, ideally, are flat crystal faces (see section 3.2.1.1) or cleavage planes (see section 3.2.3.3). Weathered surfaces will usually result in an underestimation of a mineral's hardness. Second, after doing a scratch test, there will be a line of powdery residue. Depending on their relative hardnesses, this could come from either the unknown mineral or the index mineral. Check the former by feeling, with a fingernail, for a scratch-induced groove or by brushing away the residue and looking at the surface with a hand-lens. If there is no groove, then the powdery residue came from the index mineral. Third, be sure to apply a firm, steady pressure while trying to scratch one mineral with another, but do not push or pull so hard that you break one of the minerals. Fourth, the scratch test works best when the unknown mineral is a single crystal or crystal fragment. If it is a multi-granular, monomineralic aggregate, be sure that you are scratching the grain interiors rather than merely breaking the grains apart along their weaker boundaries. This is especially a problem for the fibrous and foliated aggregates. And fifth, multi-granular mineral aggregates are commonly impure, and so the hardness test will not be accurate if other minerals are present in significant amounts.

Sets of Mohs hardness index minerals (minus the diamond) can be purchased from numerous online vendors or you can create your own by buying the individual minerals. Everything, of course, is available online, but these materials might also be found locally at a 'rock shop' or the occasional 'rock and mineral show'. Although you are unlikely to ever need a diamond for mineral testing, reasonably priced diamond-tipped scribers are available. Scratching one mineral with another becomes difficult when there are no sharp corners or edges. To avoid this problem, some vendors of Mohs hardness testing kits offer stylus-like picks with tips cut from the index minerals. The drawback with these, however, is that the narrow tips break off too easily, especially those for the softer minerals. A better, although more expensive, approach is to replace the index minerals on the picks with metal tips of equivalent hardness. I am currently aware of only one company that manufactures such a set, Mineralab (*www.mineralab.com*). This provides, as part of the set, a small piece of silicon carbide for sharpening the tips, which range from 2 to 9 in hardness. Also, the metal tips can be easily removed and inexpensively replaced. The only problem is that Mineralab does not provide documentation to purchasers (or to me when I requested it) specifying how closely the scratch hardness of the metals match the index minerals that they represent. Buyers simply have to trust that the pick hardnesses are correct. Judging from the global popularity of these tools and the apparent lack of complaint about their accuracy, it appears that they are a good alternative to using the actual index minerals.

Tests of scratch hardness can also be done with inexpensive, commonplace materials. The ones usually employed, and the only ones needed for the determinative tables in this book, are a fingernail, a piece of metallic copper, a steel knife blade, and a piece of plate glass.

Fingernail: an unpainted and otherwise unadorned human fingernail has a Mohs hardness of 2 to 2.5.

Metallic copper: relatively pure copper and also bronze (copper with up to several percent tin) have a Mohs hardness of 3 to 3.5. Oddly, virtually every recent North American or British mineralogy reference suggests using a 'copper (or bronze) penny' to test minerals, even though such coins have not been minted for many decades now. The US and British pennies (and Euro one cent) are now made from steel or zinc with a thin electroplated coating of copper. The coating might not be thick enough for hardness tests but there are better sources of copper you can use. For example, inexpensive pipe fittings of pure copper are available wherever plumbing supplies are sold.

Knife blade: depending on the mineralogy reference book, the Mohs hardness of a steel knife blade is reported as anything between 5 and 6.5. This is not surprising given the many varieties of steel that are used for blades. The full range of scratch hardnesses for commercially available steel alloys is 4 to 8, and so knife blades fall in the middle. The most commonly reported blade hardness, and the one consistently found in my limited testing of knife blades, is 5 to 5.5. You cannot assume, however, that this is the hardness of your knife. I strongly recommend that you put together a set of three of the Mohs index minerals – crystals of apatite (5), orthoclase (6), and quartz (7) – and use these to determine the hardness of your blade. You do not have to use a knife, however. Possible alternatives are steel dental picks, nails, and screwdrivers, but these also need to have their hardness determined prior to use. In my experience, dental picks are always over 5 in Mohs hardness, but nails (except for the hardened, fluted ones used for concrete masonry) and screwdrivers are mostly less than 5.

Plate glass: as with steel knife blades, a range of Mohs hardnesses have been reported in the mineralogical literature for common 'plate' (or 'flat') glass used in windows, but the range is a little narrower: 5.5 to 6.5 with the great majority of sources reporting it as either 5.5 or 5.5 to 6. This is completely in accord with my own limited testing. I also found the same 5.5-6 hardness for Pyrex laboratory vessels, glass food jars and bottles, and drinking glasses. Nevertheless, whatever glass is used for Mohs hardness testing, its hardness first needs to be determined with the 5 to 7 index minerals.

The determinative tables in most mineralogical references, and also those in the present book (Tables 3.4-3.6), separate minerals into three Mohs hardness groups: ≤ 2.5, > 2.5 to ≤ 5.5, and > 5.5. The separations can be done more-or-less correctly with the above commonplace materials, but one could also employ minerals that have well-established hardness values of exactly 2.5 (e.g. brucite, calaverite, chalcanthite, or cryolite) and 5.5 (cobaltite, perovskite, or willemite). These can be purchased from many mineral supply companies.

3.2.3.2 Tenacity

Tenacity is a mineral's resistance to breakage. It is best observed in narrow splinters and thin flakes. There are nine categories of this property.

FOR CRYSTALS AND CRYSTAL FRAGMENTS

Brittle: breaks easily into small pieces. In other words, the mineral has little or no resistance to breakage when struck with a hammer. This is the kind of tenacity exhibited by the great majority of minerals.

Sectile: slices can be cut off with a steel blade and these will either crumble or, if malleable, flatten into a sheet when hammered.

Malleable: can be easily flattened into thin sheets by hammering without breaking into fragments. Such minerals are also always sectile.

Ductile: can be drawn into the form of a wire. Such minerals are also always both malleable and sectile.

Flexible (or **Pliable**): a thin piece will bend without breaking and remain bent after the deforming force is removed.

Elastic: a thin piece will bend without breaking but will spring back to more-or-less its original shape after the deforming force is removed.

FOR AGGREGATES

Tough: very difficult to break, even under hammering.

Friable: small pieces are readily dislodged by rubbing with a finger.

Pulverulent: easily reduced to powder by applying pressure.

3.2.3.3 Cleavage

Cleavage is the tendency of minerals to break consistently and repeatedly in certain directions. These are the planes of weakness within a mineral and are an outward manifestation of the symmetry of the three-dimensional arrangement (i.e. structural lattice) of its constituent atoms. Not every mineral exhibits cleavage, but those that do will have one of the following types as determined by their crystal system (Figure 3.8).

Basal or **Pinacoidal**: breakage in only one direction that produces sheet-like or platy fragments. Basal (a.k.a. 'basal pinacoidal') cleavage occurs only in the tetragonal and hexagonal crystal systems, and is perpendicular (i.e. intersects at a 90° angle) to the crystallographic 'c' axis (Figure 3.1) whereas ordinary pinacoidal cleavage is found only in the orthorhombic, monoclinic, and triclinic crystal systems and has orientations that are oblique to the 'c' axis.

Prismatic: breakage in two directions intersecting at or near a 90° angle in the tetragonal crystal system or at an oblique angle in the hexagonal, monoclinic, and orthorhombic crystal systems. All cleavage directions are parallel to the 'c' axis. The fragments produced tend to be long to mainly short and blocky prisms.

Cubic: breakage in three directions intersecting at 90° and occurring only in the isometric crystal system. The fragments produced tend to be six-sided with a cubical shape.

Rhombohedral: breakage in three directions intersecting at oblique angles and occurring only in the hexagonal crystal system. The fragments produced can, under controlled breakage, be six-sided with the shape of a rhombohedron (i.e. a rhomb).

Octahedral: breakage in four directions intersecting at oblique angles and occurring only in the isometric crystal system. The fragments produced can, under controlled breakage, be eight-sided with the shape of an octahedron.

Dodecahedral: breakage in six directions intersecting at oblique angles and occurring only in the isometric crystal system. The fragments produced can, under controlled breakage, be twelve-sided with the shape of a dodecahedron.

Most cleavable minerals have just one type of cleavage, but some have either both basal/pinacoidal and prismatic cleavages, or two or more pinacoidal cleavages with different orientations.

The names for the types of cleavage are the same ones given to some of the general crystal forms in Figure 3.2, but this is coincidental. The forms of both cleavage fragments and crystals are determined by a mineral's internal structure, but it is the planes of weakness producing the cleavage directions and the planes with the highest density of atoms that are responsible for the crystal faces.

Cleavage is a powerful diagnostic tool for mineral identification but only in the hands of a skilled mineralogist. It is usually easy to recognize that cleavage of some kind exists, provided that it is well developed, but determining the number of directions and their angular relationships will be difficult to impossible for most people. There are two reasons for this. First, cleavage will only be visible in crystal fragments with broken surfaces that are large enough for the cleavage planes to be clearly seen with either the naked eye or under a hand-lens. Even then, not all cleavage directions will be obvious. Cleavage will not be visible at all when a mineral occurs as either an unbroken crystal or a fine-grained aggregate. In the latter case, the mineral grains are just too small for any cleavage surfaces to be visible. In crystals that have developed internal cracks, however, these breaks will usually follow cleavage directions. And second, some cleavage directions in a given mineral will be better developed than others. Mineralogists employ the following subjective terminology to describe the quality of a cleavage direction: **perfect** (or **eminent**) – it is hard to break the mineral in any other direction; **distinct** – breakage is mainly along the cleavage and usually results in smooth, flat surfaces; **good** – breakage occurs easily along the cleavage directions, although in less continuous and somewhat irregular surfaces, but the mineral also readily breaks in other directions; and **fair** to **poor** – cleavage is poorly developed and difficult to see. For Tables 3.4-

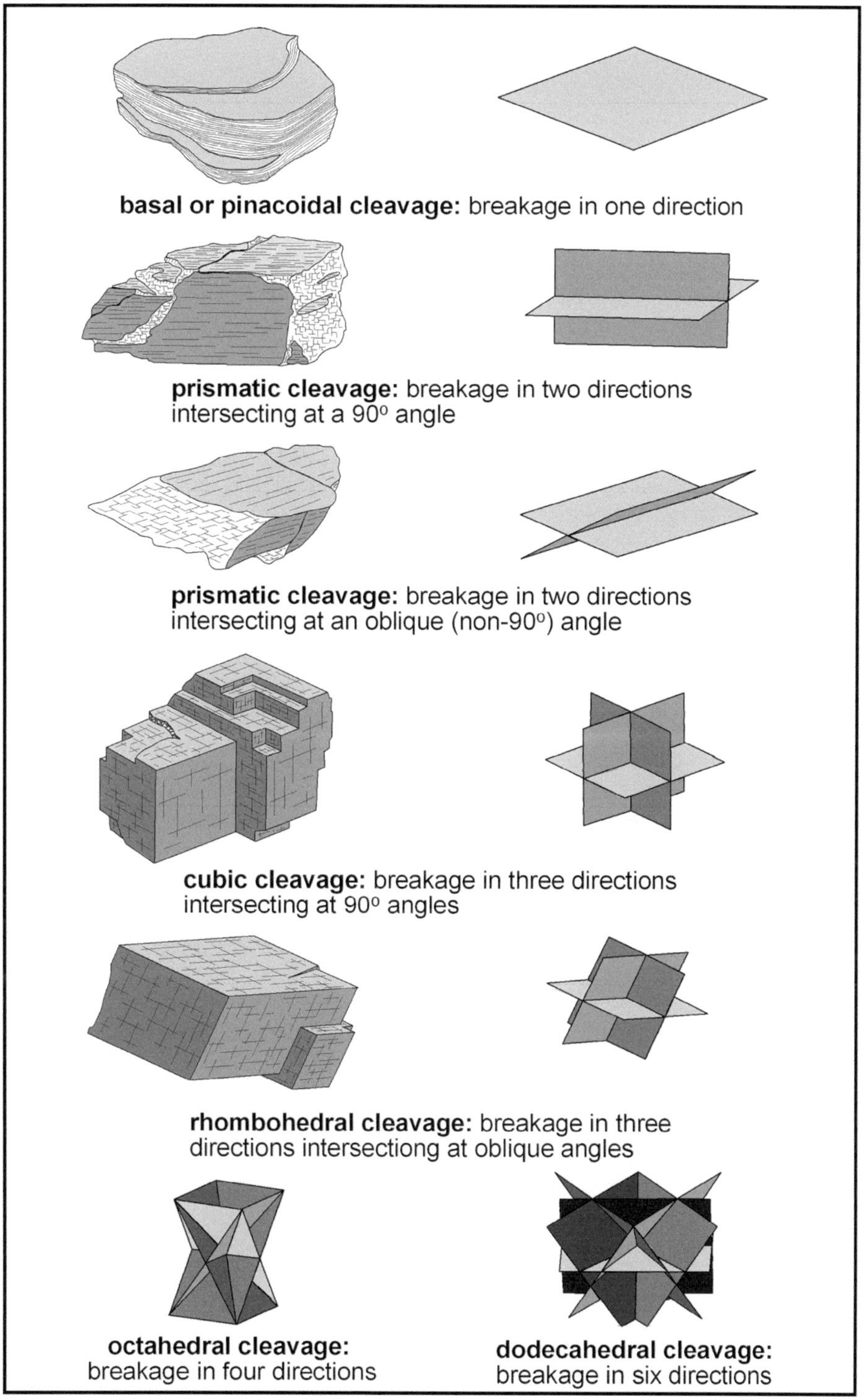

Figure 3.8: Types of cleavage possible in minerals.

3.6, cleavage quality is simplified to just two levels: **prominent** when perfect or distinct, or **not prominent** when good, fair, or poor.

It is sometimes difficult to tell whether a flat surface on a mineral specimen is a cleavage plane, a crystal face (see section 3.2.1.1), a parting plane (see section 3.2.3.4), or simply a surface that happens to be flat due to an even fracture (see section 3.2.3.5). Cleavage planes repeat themselves at different depths within a mineral and so on a broken surface the cleavage in a given direction will usually manifest itself as a set of parallel, slightly offset, flat surfaces (like a deck of cards that has been splayed to one side). This will not be the case for crystal faces and even fractures, but parting can have a similar appearance. The best way to see cleavage surfaces is under a bright light with a hand-lens. Rotating the mineral in the light will cause the highly reflective cleavage surfaces to sparkle. Those for a given cleavage direction will sparkle together and further rotation will cause the surfaces for another cleavage direction, if present and well developed, to light up simultaneously.

A further source of difficulty and confusion in analyzing cleavage is the fact that the standard mineralogical references commonly disagree on the type and quality of a mineral's cleavage, especially when it is not well developed. For the sake of consistency, therefore, the cleavages reported in Tables 3.4-3.6 come from the comprehensive listing of Sinkankas (1972: 186-193). Those given for minerals not included in Sinkankas' list represent the consensus opinion among Berry et al. (1983), Gribble (1988), Klein and Dutrow (2007), and *mindat.org*.

3.2.3.4 Parting

Parting is a rare phenomenon that is similar to cleavage in that it is breakage along internal planes of structural weakness. Parting surfaces, however, tend to be discontinuous and rough, and present only on some specimens of a given mineral. Usually when a specimen exhibits parting it is a response to some kind of directed stress experienced during the mineral's entombment within its source rock. When present, parting may be indistinguishable from cleavage.

3.2.3.5 Fracture

Minerals that lack cleavage will break by fracturing. This is also true of minerals with poorly developed cleavage. The fracture surfaces produced by natural causes (i.e. not by intentional knapping) are randomly oriented and do not occur in parallel sets. There are five types of fracture surfaces (Figure 3.9).

Conchoidal: a curving surface (either concave or convex) with smooth concentric ridges diminishing in radius toward the point of initial breakage.

Even: a smooth, nearly flat and planar surface.

Uneven: a rough, irregular surface.

Hackly: a ragged surface with sharp, jagged edges and points.

Splintery: a surface with partially separated splinters or fibers.

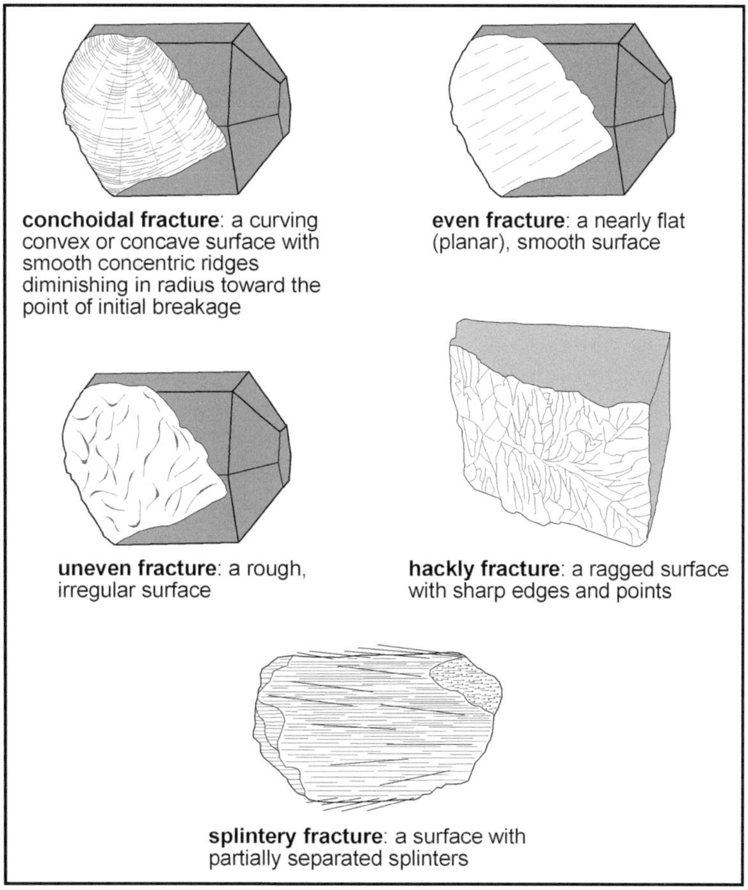

Figure 3.9: Types of fracture possible in minerals.

3.2.4 Density and Specific Gravity

3.2.4.1 Definitions and Overview

The 'density' of a mineral specimen is defined as its weight divided by its volume. This is usually expressed as grams per cubic centimeter (g/cc). A specimen's weight, and therefore also its density, decreases slightly with increasing elevation above sea level and this is due to a progressive decrease in the gravitational constant with increasing distance from the Earth's center. For example, a mineral specimen weighing 100.00 g at sea level will weigh 99.91 g at 3000 m above sea level or 0.09% less. If the mineral is, say, quartz with a density of 2.651 g/cc at sea level, then it would have a density of 2.648 g/cc (0.09% less) at 3000 m. After rounding off to the nearest hundredth, however, the numerical value of density remains unchanged. You would have to be above about 6000 m when measuring density for the after round off value to differ in the second decimal place. Thus, for all practical purposes, elevation can be ignored when measuring density.

The 'specific gravity' of a mineral specimen is the ratio of its weight to the weight of an equal volume of pure water at 4°C. This is equivalent to the ratio of the density of the mineral to the density of the water at 4°C and, consequently, specific gravity is also known as 'relative density'. Specific gravity is a dimensionless number because the weight (or density) units cancel out. A ratio of weights is unaffected by elevation but the temperature of the water does matter. Water is at its minimum volume at 4°C for a given mass and, therefore, is also at its maximum density at this temperature. Water density decreases at both higher and lower temperatures. For example, it decreases from 1.0000 g/cc at 4°C to 0.9922 g/cc at 40°C. Again using the example of quartz, this mineral's specific gravity will vary, due to the effect of temperature on water, from 2.651 g/cc at 4°C to 2.649 g/cc at 10°C, 2.645 g/cc at 20°C, 2.639 g/cc at 30°C, and 2.630 g/cc at 40°C. Although the temperature effect is small, it is significant in the second decimal place. It can, nevertheless, be ignored in mineral identification because an approximate specific gravity value is sufficient given all the other macroscopic properties available to assist in the identification process. Additionally, the techniques described below for determining density and specific gravity will only yield approximate values when the measurements are made under field conditions with less than optimal equipment and mineral specimens.

Density and specific gravity have the same numerical value at 4°C and sea level. Consequently, the two measures are used interchangeably in the mineralogical literature and can be considered as essentially equivalent. The reason for this equivalence is that in the metric system of measurement, one gram is defined as the weight of one cubic centimeter (which is equal to one milliliter) of pure water with this originally defined as the weight at the melting temperature of ice or 0°C, but later changed to 4°C when it was discovered that water is at its most dense at this temperature. It is always the density/specific gravity at 4°C that is reported in the literature and also in the determinative tables in this book. Some minerals, like quartz (SiO_2), have an invariable chemical composition and so their specific gravity has a fixed value (2.6508 for quartz but almost always reported as simply 2.65). Other minerals, like garnet ($\alpha_3\beta_2[SiO_4]_3$ where α can be Ca, Fe, Mg and/or Mn, and β can be Al, Cr, and/or Fe), are a solid-solution series where the specific gravity varies with composition (3.1 to 4.3 for garnet). Still other minerals have a fixed chemical formula, like galena (PbS), but commonly have abundant elemental impurities that result in a variable specific gravity (7.4 to 7.6 for galena).

A mineral's specific gravity (or density) is potentially the most useful of its diagnostic properties. Although reported in this book's determinative tables (Tables 3.4-3.6), it is not used in organizing them, except at the lowest level, because its measurement is often not possible. This would be the case if the necessary equipment is not available or if a specimen is not monomineralic, is too small or too large for the equipment, reacts with the water used in the measurement (e.g. dissolves or swells), or contains significant internal porosity with trapped air (e.g. along grain boundaries in aggregates or along fractures and cleavage cracks in crystals). The measurement process would not be a problem, however, for water-soluble minerals (like halite) if the immersion lasts no more than 5-10 seconds as would be the case for the methods described in section 3.2.4.3. Only an insignificant amount of material can be dissolved in that time. If you are concerned that you might be dealing with a water-soluble mineral, there is a simple test for recognizing this property: taste (see also section 3.2.10). Because such minerals will dissolve in your saliva, they will have a strong taste as reported in Table 3.6. There is a 'dry' alternative to the water-based methods described below for finding the volume of a specimen, but it will seldom be available to archaeologists. This is 'helium pycnometry', which replaces water with a chemically inert gas: helium. The equipment is expensive and will only occasionally be found in engineering and science laboratories.

Mineral specimens being analyzed should obviously be as free as possible of any adhering dirt or other encrustations. If it is porous, or a mixture of minerals, then it should first be crushed to a sandy grit both to open the pores and to allow the analyst to pick out the contaminating mineral grains. Elemental impurities in a mineral are normally not a problem because the common ones are factored into the specific gravity ranges reported in Tables 3.4-3.6.

Once you have some experience with minerals of different specific gravity, it becomes possible to get a general, but still diagnostically useful, idea of this property's value just by the 'heft' of a specimen. This is done with the unknown mineral in one hand and a known mineral of similar size in the other. It would be useful for this purpose to keep on hand several pieces of different sizes of a common mineral like quartz or calcite (or calcite-rich rocks like limestone or marble) because the average specific gravities of these minerals (2.65 and 2.71, respectively) fall in the mid-range of the spectrum of mineral specific gravities. The idea of a mid-range specific gravity is also useful for dividing the minerals in Tables 3.4-3.6, at their lowest level of separation, into two specific gravity ranges: ≤ 2.70 and > 2.70. The minerals in these tables, along with a large number of other geologically common and economically important minerals, are listed in order of increasing specific gravity in Table 3.9.

3.2.4.2 Equipment and Other Considerations

Balances and Scales: an electronic balance is best for measuring specific gravity provided it is accurate to at least the nearest hundredth of a gram. Such an instrument will commonly have a hook on the underside for weighing suspended loads, but if yours does not, you can still make the measurement with the top-loading pan. A less expensive, but also slightly less precise, alternative to an electronic balance is a mechanical spring scale. The better-quality spring scales can only reproducibly weigh a load to the nearest tenth of a gram, which is still good enough for an approximate specific gravity. For this level of precision to be possible, however, a given spring will have a very limited load range in terms of its maximum capacity and so two or more overlapping spring scales may be needed to cover the same load range of the typical electronic balance. One can also, of course, use a mechanical balance-beam scale but such an old-fashioned instrument is hard to find nowadays and would, in any case, be less precise than a good electronic balance.

Water and Temperature: ideally it is pure (i.e. distilled or deionized) water at 4°C that is used in the measurement process but, in practice, ordinary tap or bottled drinking water at any temperature works about as well. Any effects on water density caused by departures from the ideal composition and temperature will be vanishingly small in comparison to the other potential sources of measurement error.

Thread: for some measurement methods, the mineral specimen needs to be suspended in water by a thread. Because this is being weighed along with the specimen, it is important that the thread's weight be negligible in comparison to that of the specimen and also that the thread neither displaces a significant amount of water nor absorbs it. The thinnest monofilament nylon thread one can find will work best, and if this cannot be found in a shop with sewing supplies, then a highly flexible fishing line will also serve. Usually this will be colorless, making it almost invisible, but if a colored thread can be obtained, it will be easier to work with.

Bubbles: the weight determination for a mineral suspended in water will be adversely affected if lots of air bubbles adhere to the thread or specimen (i.e. the weight will be underestimated). Tapping the thread or specimen or gently sweeping them with a brush will dislodge most of the bubbles. You can completely avoid them, however, by first boiling the water and then allowing it to cool before using. Alternatively, and nearly as effective, add 'one drop' of a wetting agent, like ordinary household (e.g. dish) detergent, to the water and gently stir. This amount of detergent is not enough to significantly affect the water's density, but it is sufficient to reduce its surface tension and so makes it harder for air bubbles to adhere.

Accuracy: the two techniques employing direct weighing in air and water (section 3.2.4.3) are the most accurate ways to measure specific gravity. When conducted properly and using an electronic balance, an accuracy of +/- 0.01 g/cc is possible, which is well within the natural variation of most minerals. The two water-displacement techniques (section 3.2.4.4), however, only work well with large specimens because for smaller ones the associated measurement errors are large in comparison to the small water displacements. In my own experiments, I only got consistently good results for specimens over about 25 cc in volume, but even then the specific gravity values for a quartz specimen fluctuated by up to +/- 0.05 g/cc. For less voluminous specimens, the measured values can be off by +/- 0.5 g/cc or more.

3.2.4.3 Measurement of Specific Gravity by Direct Weighing in Air and Water

USING EITHER AN ELECTRONIC BALANCE WITH A BOTTOM-LOADING HOOK OR A MECHANICAL SPRING SCALE

Step 1: weigh the mineral specimen in air while suspended by a fine thread attached to the hook on the underside of a balance or spring scale (Figure 3.10a). Record the weight (Wt_{air}).

Step 2: position a beaker of water so that the suspended specimen is completely immersed while not touching the beaker's side or bottom (Figure 3.10b). Record the weight of the submerged specimen (Wt_{water}). The numerical difference between a specimen's weight in water and in air equals the weight of water with the same volume as that of the specimen.

Step 3: specific gravity = $Wt_{air} / (Wt_{air} - Wt_{water})$
= weight of specimen / weight of an equal volume of water

USING A TOP-LOADING ELECTRONIC BALANCE

Step 1: weigh the mineral specimen on the top pan ($Wt_{specimen}$; Figure 3.11a).

Step 2: place a beaker partially filled with water on the pan and weigh it (Wt_{water1}; Figure 3.11b).

Step 3: submerge the specimen in the water while suspending it from a thread (Figure 3.11c). Take care that the specimen does not rest against the beaker's side or touch its

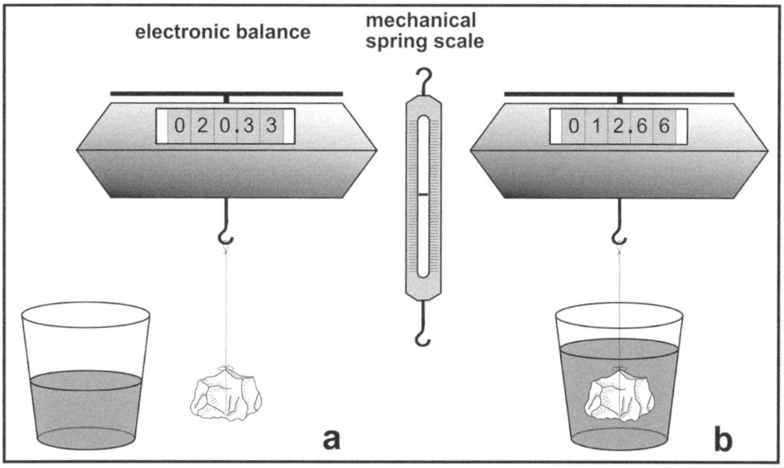

Figure 3.10: Measurement of specific gravity using either an electronic balance with a bottom-loading hook or mechanical spring scale. Numbers are for quartz with a specific gravity of 2.65.

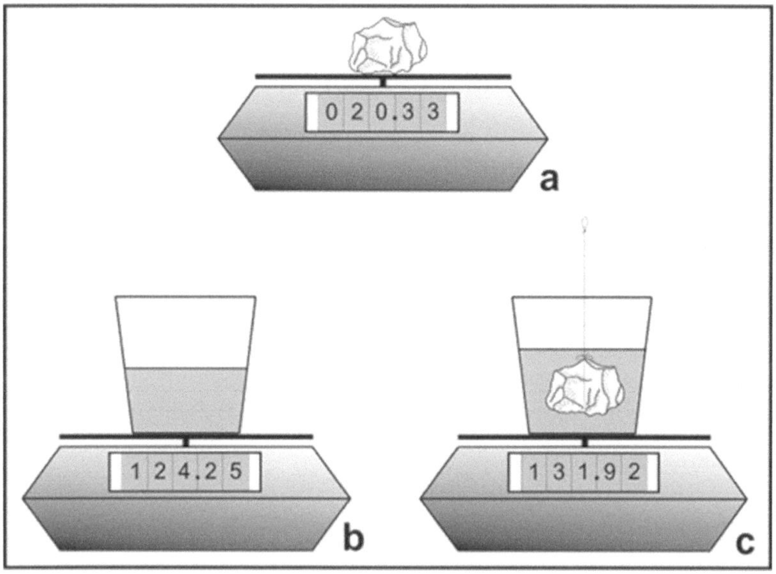

Figure 3.11: Measurement of specific gravity using a top-loading electronic balance. Numbers are for quartz with a specific gravity of 2.65.

bottom. Record the weight (Wt_{water2}). The increase in weight from step 2 to 3 corresponds to the weight of the water with a volume equivalent to that of the specimen.

Note: the reader may be surprised that the balance reading is affected by the specimen, given that it does not touch the beaker and so, seemingly, does not 'weigh' on the scale. Archimedes' Principle is at play here. According to this, an object wholly submerged in water is buoyed by an upward force equal to the weight of the displaced water, with the latter a downward-directed force that is manifested as an increase in the total weight recorded by the balance.

Step 4: specific gravity = $Wt_{specimen} / (Wt_{water2} - Wt_{water1})$
= weight of specimen / weight of an equal volume of water

3.2.4.4 Measurement of Density and Specific Gravity by Direct Weighting in Air and Water Displacement

Using a graduated cylinder

Step 1: weigh the mineral specimen with either an electronic balance or mechanical spring scale (Figure 3.12a). Record this weight ($Wt_{specimen}$).

Step 2: partially fill with water a cylinder (tall beaker) that is graduated in milliliters or multiples thereof (Figure 3.12b). Record this level (Vol_1).

Note: the more finely graduated the cylinder, the better the result (assuming, of course, that the graduations are accurate). The water level should be set precisely at a milliliter marking, using either the bottom or the easier to use top of the water's concave-downward meniscus.

Step 3: immerse the specimen in the water either by dropping it in (without splashing) or suspending it from a thread (Figure 3.12c). Record the new water level (Vol_2). The difference between Vol_1 and Vol_2 corresponds to the specimen's volume in milliliters or, equivalently, cubic centimeters. This difference also represents the weight (in grams) of water with a volume equal to that of the specimen (Wt_{water}).

Step 4: specific gravity = $Wt_{specimen}/Wt_{water}$
= weight of specimen/weight of an equal volume of water
density = $Wt_{specimen} / (Vol_2 - Vol_1)$
= weight of specimen / volume of specimen

Using a beaker with an overflow spout

Step 1: weigh the mineral specimen with either an electronic balance or mechanical spring scale (Figure 3.13a). Record this weight ($Wt_{specimen}$).

Step 2: weigh an empty beaker (Figure 3.13b; Wt_{beaker}).

Step 3: add enough water to a beaker with an overflow spout to initiate flow out of the spout. Wait for the water level to stabilize within the beaker. This level will now be at the bottom edge the spout. Place the empty beaker from Step 2 under the spout (Figure 3.13c).

Note 1: metal, plastic, and Pyrex glass beakers with an overflow spout are available from catalogs of classroom supplies for the Earth Sciences and Physics.

Step 4: either drop the specimen into the water-filled beaker (without splashing) or immerse it while suspended from a thread (Figure 3.13d). The displaced water flows into the empty beaker under the spout. Before removing the beaker for weighing, wait for the water to stop dripping.

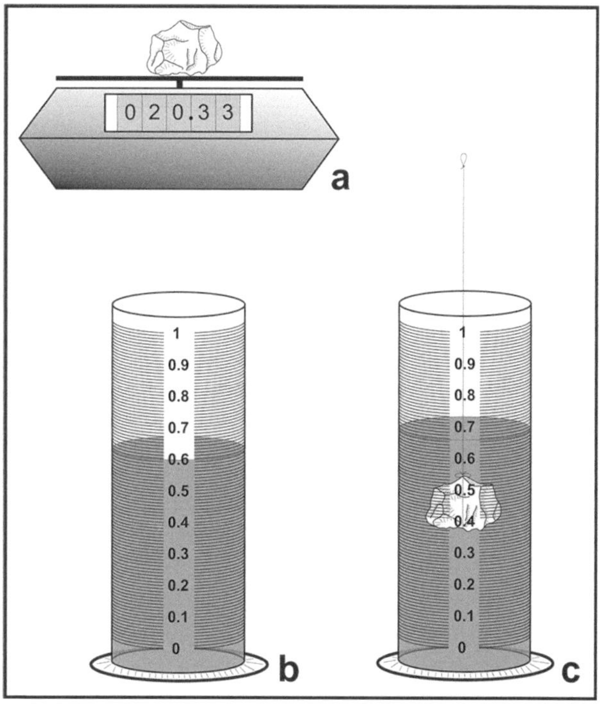

Figure 3.12: Measurement of density and specific gravity by direct weighing in air and water displacement in a graduated cylinder.

Note 1: during this step, it is essential that the spout be angled downward to prevent the overflowing water from running out along the bottom of the spout and then down the side of the beaker. If the spout does not already have this angle, place something under the opposite side of the beaker to tilt the spout. Alternatively, attach a length of water-tight tubing to the spout to carry the overflow directly into the beaker. Once the water stops flowing, tap the spout or tubing to dislodge any adhering water droplets.

Note 2: the water's surface tension will inhibit its free flow through the spout, and so to reduce this effect add a few drops of a household detergent and stir this into the water in the beaker.

Step 5: weigh the beaker with the accumulated water from Step 4 ($Wt_{beaker+water}$; Figure 3.13e). The difference between this weight and that of the empty beaker in step 2 is the weight of the displaced water, which corresponds to both the weight of the water with the same volume as the specimen and the volume of the specimen ($Vol_{specimen}$).

Step 6: specific gravity = $Wt_{specimen} / (Wt_{beaker+water} - Wt_{beaker})$
= weight of specimen / weight of an equal volume of water

density = $Wt_{specimen} / Vol_{specimen}$
= weight of specimen / volume of specimen

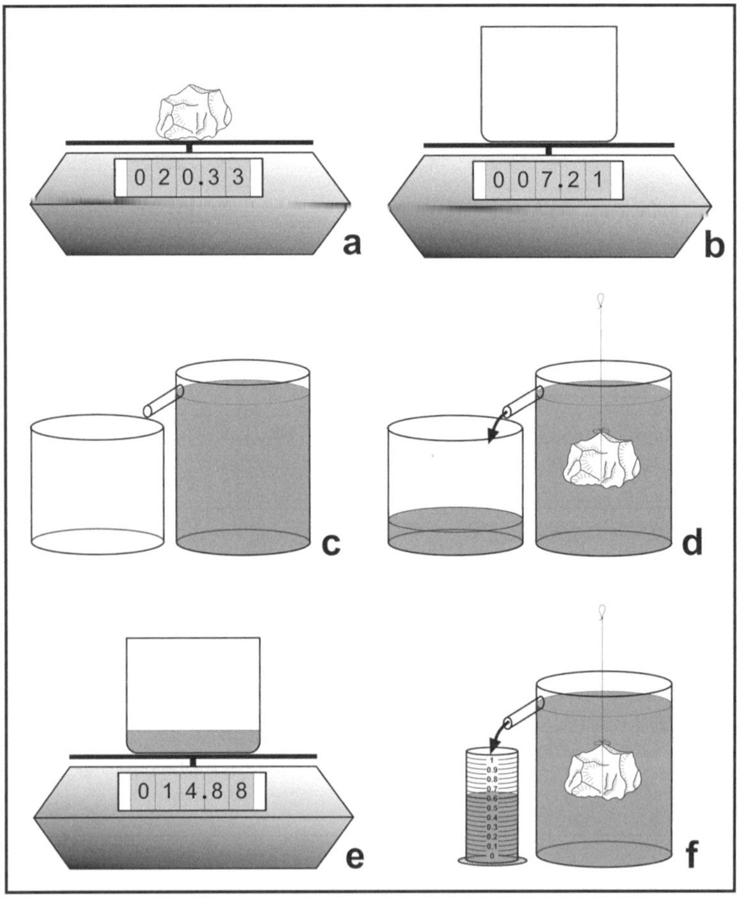

Figure 3.13: Measurement of density and specific gravity by direct weighing in air and water displacement in a beaker with an overflow spout. Numbers are for quartz with a specific gravity of 2.65.

An alternative means of determining the density is to direct the overflowing water into a graduated cylinder (Figure 3.13f), where the water volume, and hence also the specimen volume ($Vol_{specimen}$), can be read off the side of the cylinder.

3.2.5 Magnetism

Several minerals are noticeably attracted to a magnet with their responses varying from weak to strong. Most of these are rich in elemental iron, which is the source of the mineral's magnetism. Sometimes other elements, such as manganese, can cause minerals to be magnetic. It is iron, however, that is responsible for the more strongly magnetic minerals and the strongest of these is the appropriately named magnetite (Fe_3O_4).

The strength of a magnetic attraction varies with the extent to which the unpaired electrons in atoms of iron (or other element) spin in the same direction: the greater the proportion spinning together, the stronger the magnetic field produced with the spin direction determining the polarity (i.e. which sides of an atom have north- and south-seeking magnetic poles). The spins are mostly aligned in magnetite but in some specimens of this mineral the alignment is so well developed that the specimen is not merely strongly attracted to a magnet but becomes a magnet itself. This is the case for the variety of magnetite known as 'lodestone', which is created by the strong magnetic fields associated with lightning strikes. Heating a weakly magnetic mineral in an open-air flame sometimes allows the electron spins to align to some degree with the Earth's magnetic field, and upon cooling the mineral will be more strongly magnetic than it was before. Interestingly, any iron-based metal, such as a knife blade or sewing needle, can be made into a magnet (i.e. magnetized by 'magnetic induction'). This is done by rubbing it with a magnet at least several times in the same direction. Any type of magnet can be used, but the weaker its magnetic field, the more times it has to be rubbed along the blade or needle. This is probably how magnetized needles were made in antiquity for navigation: using a natural lodestone magnet to induce the magnetic field. The lodestone itself, if suspended from a string or placed on a floating piece of wood, will also act like a compass.

Both normal and post-heating magnetism are useful diagnostic properties for several minerals and it can be tested with a small, inexpensive magnet. The most common magnets are iron-based and now consist of 'ferrite', a ceramic composed mainly of iron oxide. Before the mid 1900s, however, they were made of a metallic iron alloy. The so-called 'rare earth' magnets, introduced in the late 1900s, are up to several times stronger than a ferrite magnet of the same size. The most powerful of these are made from an alloy of neodymium or Nd (a rare-earth element; atomic number 60 in Table 2.1) together with iron and boron. Although these cost a little more than the ferrite magnets, they are still relatively inexpensive and widely available. They come in a wide variety of shapes and sizes, but a cylindrical rod-like form is the most useful for mineral testing if this is to be suspended from a thread (see below). The larger the magnet, the greater its magnetic field strength, or 'pull force', and hence the greater its sensitivity to weakly magnetic minerals. There is no one right size or strength that can be recommended because there is no testing standard to adhere to, but certainly you only need a relatively small rare-earth magnet. In my own work, I use a neodymium magnet that is a cylindrical rod 12 mm long by 5 mm in diameter with a pull force (i.e. the weight required to pull it vertically down and off the underside of an iron plate) of 3.4 kg. This is entirely satisfactory for testing minerals (and rocks). I recommend that you bring at least a few of these to the field or lab because, being so small, they are easily lost. A useful alternative is to purchase a scriber with a silicon or tungsten carbide (or diamond) tip and a rare-earth magnet at the other end.

Determining the strength of a mineral's magnetic attraction is not as straightforward as you might suppose. The observed strength of attraction depends very much on the field strength of the magnet. Magnetic attraction also increases slightly with the size (mass) of the mineral specimen and the flatness of its side facing the magnet. Minerals exhibiting a magnetic attraction are described in the mineralogical literature (and in Tables 3.4-3.6) in relative terms like 'strong' or 'weak', and sometimes also 'moderate'. How these determinations were arrived at is never stated, but they probably involve a variety of iron-based magnets with a range of strengths. Nevertheless, a strong magnetic attraction is fairly unambiguous and easy to demonstrate. If a magnet of any kind clings tightly and unsupported to the underside of a mineral specimen, then it is strongly attracted and this finding would be consistent with any earlier claims of strong attraction for this mineral in the literature. Minerals that do not behave in this way have either moderate, weak, or no magnetism. Distinguishing among the latter outcomes is problematical because their recognition is both subjective and very much dependent on the magnet being used. Even so, the test is still worth conducting. One holds a magnet against the specimen and then slowly pulls it away while sensing the amount of pull force. It is on this basis that you decide whether the attraction is moderate, weak, or non existent. A variation on this test, which makes it more sensitive, is to suspend a rod-shaped magnet from a thin string (or thread), which is secured around the magnet's mid-section, and then observe the degree to which the attraction pulls the magnet toward the specimen and the string out of the vertical.

Magnets are also useful to archaeologists for identifying manufactured iron, which is strongly magnetic, or for extracting small pieces of this metal from a soil or sediment. In the latter case, you will find it advantageous to place a sheet of thin paper between the loose soil/sediment and magnet. In this way, the iron particles can be easily removed from the underside of the paper by pulling the magnet away. Without the paper, you will find it surprisingly difficult to remove all the attached particles from the magnet. Heat-induced magnetism usually exists in the burnt (and reddened) sediment, soil or rock under ancient fire pits for cooking and heating, kilns for brick and pottery manufacture, furnaces for smelting metals, and forges for smithing metals. The magnetism, which is associated with iron oxides, is very weak. Although a hand magnet cannot detect them, such 'magnetic anomalies' are easily identified with a highly sensitive instrument known as a magnetometer.

3.2.6 Induced Electrical Charge

Many minerals develop a temporary electrical charge when vigorously rubbed with a wool or silk cloth. An old wool sock works well, but the test requires low-humidity conditions. The presence of the induced charge in a mineral specimen, called **frictional electricity**, is demonstrated by holding it (preferably with rubber gloves or other insulating material) close to small, light-weight objects such tiny bits of paper, tuffs of cotton, pieces of thread, or the hair on your arms or head. A few minerals can acquire an electrical charge by very different means; either through applying pressure (either tensional or compressive stress) in certain directions (**piezoelectricity**) or by heating (**pyroelectricity**).

3.2.7 Odor and Feel

There are three distinctive odors produced by either heating or freshly crushing certain minerals or rocks: (1) **garlicky** – smells like garlic and is characteristic of arsenic-bearing minerals; (2) **sulfurous** – smells like a freshly lit match (due to sulfur dioxide, SO_2, fumes) and is characteristic of sulfur-bearing minerals; and (3) **bituminous** – smells like hot asphalt or tar and is characteristic of bitumen (hydrocarbon)-bearing rocks. Heating some minerals can also produce what is commonly described as a **burnt** odor. An **argillaceous** or **musty** odor, with a smell like wet modelling clay, occurs when breathing upon (and thereby moistening) clay minerals or clay-rich rocks. A **fetid** odor, with a smell like rotten eggs (due to hydrogen sulfide, H_2S, fumes), is characteristic of some sulfur-bearing minerals when immersed in hot dilute hydrochloric acid.

Minerals can have different tactile characteristics and these are usually attributable to how they break and the surfaces thereby produced. Apart from these, however, there is a **greasy** feel that is like the soft, smooth surface of a bar of soap and this is characteristic of talc and some clay minerals. Clays and some clay-rich rocks will stick to your tongue (because they absorb your saliva) and this gives them a **dry** feel.

3.2.8 Reaction of Carbonate Minerals to Acid

Carbonate (CO_3)-bearing minerals are among the most common earth materials and are readily identified by their reaction to dilute acid: a fizzy effervescence from carbon dioxide (CO_2) gas released from the dissolving carbonate molecules. The intensity of the effervescence depends not only on the mineral species, but also on whether it is whole or powdered, and on the concentration and temperature of the acid used. The strength of an acid is indicated by its pH value, where 7 is 'neutral' and 0 to 7 is 'acid' with the acidity increasing as the pH number decreases. Solutions with a pH greater than 7 (up to 14) are said to be 'alkaline'. The pH scale is logarithmic to the base 10, and, consequently, a difference of 1 on the scale corresponds to a 10-fold difference in acidity, which is determined by the hydronium ion (H_3O^+) concentration in an aqueous (i.e. water-based) solution.

Recommended Acid Strength

Mineralogical references almost always recommend using what they refer to as '5%' or '10%' solutions of hydrochloric (i.e. hydrogen chloride or HCl) acid with acidities of, respectively, pH 1.15 and pH 0.85. Note that this is a two-fold difference in strength: i.e. 10 to the -0.85 power divided by 10 to the -1.15 power. These percentages are misleading, however, because what is actually meant is a 5 or 10 '*volume* percent' solution of standard, reagent-grade hydrochloric acid (e.g. for the 5% HCl solution, 5 ml of reagent is added to 95 ml of water for a total volume of 100 ml). The reagent, however, is not pure hydrogen chloride but rather an aqueous solution of this chemical. To manufacture it, 37-38 g of hydrogen chloride gas is dissolved in 62-63 g of pure (i.e. deionized) water for a 100 ml solution. The reagent is, thus, a 37-38 '*weight* percent' solution of hydrogen chloride. The reason for this particular concentration, which is used in all commercial HCl reagents, is that the aqueous solution releases copious amounts of hydrogen chloride gas at concentrations over about 40 weight percent. Thus, the 5 and 10 volume percent hydrochloric acid solutions actually correspond, respectively, to 2.16-2.24 and 4.29-4.44 weight percent solutions of hydrogen chloride (given a reported density of 1.18-1.19 g/ml at 20°C for the 37-38 weight percent HCl reagent). If these dilute solutions come briefly into contact with your skin, as they inevitably will, they will cause a tingling sensation but otherwise do no harm. They will, however, severely damage any clothing they touch. Fortunately, the acid is easily washed off with water.

Acid Tests

I recommend using only a 5 volume percent solution of reagent-grade hydrochloric acid (henceforth referred to simply as 5% HCl). This is concentrated enough to produce observable reactions in all the carbonate minerals. It is also less irritating than the 10% HCl if it gets on your skin. The only advantage to using 10% HCl is that it enhances the reactions, making them a little easier to see. A dilute HCl solution is primarily used to distinguish between the two most abundant carbonate minerals, which are otherwise difficult to tell apart visually: calcite ($CaCO_3$) and dolomite ($CaMg[CO_3]_2$). In a drop of 'cold' (e.g. room temperature) 5% HCl, calcite crystals and aggregates will exhibit a *strong reaction* (i.e. effervescence develops rapidly and is vigorous, and sometimes is even audible), whereas dolomite crystals and aggregates will have either *no reaction* or a *weak reaction* (i.e. effervescence develops slowly and is languid with only a few CO_2 bubbles rising through the acid drop). The dolomite, however, will react strongly with a brisk effervescence if it is first powdered, which greatly increases its reactive surface area. The test is, thus, to place a drop of acid on a whole specimen. If there is no reaction, create some powder by scratching the specimen's surface with a knife and then add another drop of acid. If there is a strong reaction, the mineral is dolomite, and if there is still no reaction, then it is probably a non-carbonate mineral. Warming the acid has the same accelerating effect on the reaction as powdering the mineral. All other carbonate minerals will react to cold 5% HCl to some degree and their response can be enhanced by the same means: either powdering the mineral or warming the acid.

Care must be taken when testing a porous aggregate. It may appear at first to have a weak reaction to the acid, but this might only be air bubbles escaping from the pores as the acid solution soaks in. If the mineral truly has a weak reaction, then this will continue long after the initial outward rush of air. Porous aggregates can be problematical in another regard: the acid may not pool on the surface and show rising CO_2 bubbles because it is continuously soaking into the aggregate. This is not a problem for minerals with a strong reaction, this will still be obvious, but weak reactions will be difficult to see. In this case, place a piece of the mineral in a test tube and add several drops of acid, enough to fully immerse it. The reaction to the acid will now be visible.

A mixture of calcite and dolomite commonly occurs in sedimentary rocks, especially in impure limestones and dolostones, but also sometimes in metamorphic marble. The minerals in such granular aggregates cannot be separated for an acid test, but there is a simple way to test the whole rock with acid that gives a general idea of the relative amounts of the two minerals and, hence, also the identity of the carbonate rock. Immerse a tiny rock chip (ideally with a platy shape) in a small, transparent vessel filled with 5% HCl and observe the reaction (Figure 3.14).

If all or nearly all calcite: the chip fizzes violently with an audible, frothy reaction; it bobs about and commonly floats to the surface, where it is buoyed up by the rising CO_2 bubbles.

If calcite exceeds dolomite: the chip undergoes a brisk but quiet fizz; it skips around the bottom of the vessel and sometimes rises slightly above it.

If dolomite exceeds calcite: the chip exhibits mild fizzing; it may rock up and down but tends to remain in one place.

If all or nearly all dolomite: there is no fizzing except for an initial slow formation of bubbles that accelerates over time to a thin stream of bubbles rising to the surface; the chip does not move.

If no bubbles form, then the chip consists largely or entirely of a non-carbonate mineral.

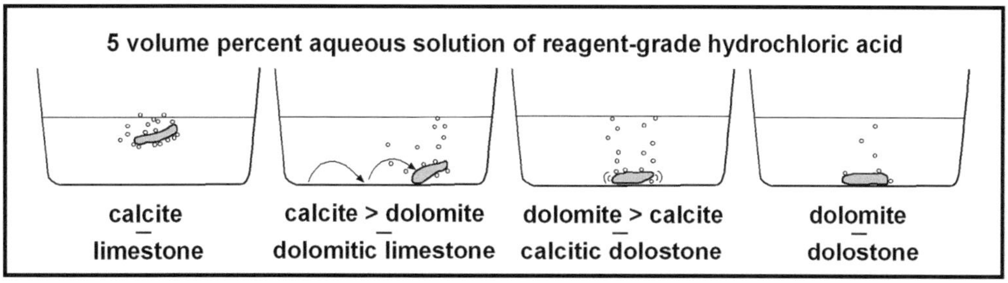

Figure 3.14: Test for relative amounts of calcite and dolomite in a carbonate aggregate. Note that these results apply not only to limestones and dolostones, but also to metamorphic marbles of different carbonate compositions.

There is another kind of reaction in hydrochloric acid that can be diagnostic for some minerals: dissolution in a hot acid solution. If you do this test, be sure you are protected from any splashing acid (see below for a discussion of safety practices when working with acids). Place a tiny amount (literally just a 'pinch') of powdered mineral in a 20 ml Pyrex beaker along with about 10 ml of either 5% or 10% HCl (collectively referred to as 'dilute' HCl in Tables 3.4-3.6). This is then heated over a candle flame until it comes to a boil. As soon as this happens, remove the beaker from the flame and set it aside to cool. A Pyrex test tube can also be used for this test, but its narrow confines can make the boiling solution somewhat explosive. After the beaker has cooled, if not before, look at the solution to see if the powder has dissolved. If it is only partially dissolved, the test is still positive for dissolution. The problem in this case is that there was too much powder for the amount of acid solution (i.e. the solution became saturated with respect to the mineral before complete dissolution could occur). Such outcomes will be less of a problem for the stronger 10% HCl. Also look at the color of the solution for the minerals that dissolved, as this can be another clue to their identity.

PREPARATION OF ACID SOLUTIONS

Dilute solutions of hydrochloric acid can be purchased online at a variety of strengths, including 5% and 10%. Alternatively, if you are at a university and it has a chemistry stockroom, the chemists there will probably be happy to prepare a solution for you. Otherwise, you will have to do it yourself from the concentrated reagent and this is a dangerous substance to handle. It is highly corrosive and so would be very damaging to your skin and, especially, your eyes, but also to your lungs if you inhale the fumes. Therefore, when handling concentrated reagent-grade hydrochloric acid, you need to rigorously adhere to the following safety practices. First, be in a well-ventilated area. A fume hood is preferable but if this is not available, then work outside in the open air or, if inside, by an open window. Second, wear rubber gloves. Ordinary dish-washing gloves are sufficient. Third, wear eye protection. A face shield is best, but plastic, wrap-around safety goggles are also good. If these are not available, then ordinary glasses will be much better than nothing. And fourth, wear thick or multi-layer clothing over your upper torso to guard against splashes. If available, a rubber apron would be best.

To prepare, say, a solution of 5% HCl (i.e. 1 part in 20 by volume) from the reagent-grade hydrochloric acid, proceed as follows. First, fill a graduated cylinder with 19 parts water (e.g. 95 ml for a 100 ml solution). Ideally this should be pure (distilled or deionized) water to avoid reactions between the acid and dissolved solids in the water. In practice, however, any reduction in acidity caused by such reactions is negligible and so it is acceptable to use impure water (e.g. tap or bottled drinking water). Second, in a separate graduated cylinder, pour in 1 part of reagent-grade hydrochloric acid (e.g. 5 ml for a 100 ml solution). Third, pour the acid into the cylinder with water. Always add acid to water and never the other way around! This prevents the generation of hydrogen chloride fumes that occurs copiously when water is added directly to the acid. Even when the mixing is done correctly, there will still be some minor fuming which is detectable as a pungent odor. Shake or stir the solution to insure uniform mixing. To prepare a solution of 10% HCl, you would add 1 part concentrated reagent to 9 parts water. Dilute acid should be stored in a Nalgene (laboratory-rated) plastic bottle with a tight-fitting cap if it is to be carried around (remember, you do not want it leaking onto your clothes or backpack). When in actual use, though, it is more convenient to have the acid in a glass bottle with a dropper cap or a plastic squeeze bottle with a sniffer tip. To test a mineral, you only need to apply one drop of the acid solution. This is most conveniently done with a dropper, but a glass or plastic rod dipped in the solution can also deliver a single drop.

Reagent-grade hydrochloric acid can only be obtained from a chemical supply company or the chemical stockroom at a university. An impure version of this acid, however, can be purchased at most hardware stores under the name of 'muriatic acid'. This comes in a wide range of concentrations, but all are less than 37-38 weight percent. Whatever the concentration, it will usually be reported on the container. Unlike the colorless reagent-grade hydrochloric acid, muriatic acid commonly contains iron impurities that give it a yellowish coloration. It is still entirely usable for testing minerals, however. To prepare a 100 ml solution of dilute muriatic acid that is equivalent in acidity (pH) to the 5% HCl described above, you need to use the 'general dilution equation', $C_1 \cdot V_1 = C_2 \cdot V_2$ where:

C_1 = concentration of hydrogen chloride in the muriatic acid (in weight percent)

C_2 = diluted concentration sought (in weight percent); taking median values from above, this will be either 2.20 or 4.36 weight percent hydrogen chloride for 5% or 10% reagent-grade hydrochloric acid solutions, respectively.

V_1 = volume of muriatic acid needed

V_2 = volume of the final solution (100 ml)

Rearranging the general dilution equation to $V_1 = (C_2 \cdot V_2)/C_1$, you can solve for V_1 provided that C_1 is known (it is usually, but not always, reported on the product label). If, for example, C_1 is 14 weight percent and you want an acidity equivalent to that of 5% HCl, then you need to add [2.20•100]/14 or 15.7 ml of muriatic acid to 84.3 ml of water. Or, more generally, you add 1 part muriatic acid to 5.37 (i.e. 84.3/15.7) parts water. To prepare the equivalent of 10% HCl from 14 weight percent muriatic acid, add [4.36•100]/14 or 31.1 ml of muriatic acid to 68.9 ml of water (or 1 part muriatic acid to 2.22 parts water).

A marginally acceptable alternative to hydrochloric acid is distilled 'white vinegar', but this is suitable only for powdered carbonate minerals. This is available in any grocery store and is also a common food ingredient (e.g. in catsup, mustard, vinaigrette and other salad dressings, sauces, canned fruit, pickles, mayonnaise, and more). White vinegar is typically sold as a 4 to 7 weight percent solution of acetic acid (i.e. 4-7 gm of pure acetic acid is added to 93-96 gm of distilled water). The percentage is usually reported on the product label and 5% is most common. Acetic acid is very weak in comparison to hydrochloric acid. Thus, 5% vinegar has a pH of 2.4, which is about 18 times less acidic than 5% HCl with a pH of 1.15 (i.e. 10 to the -1.15 power divided by 10 to the -2.4 power). Carbonate minerals will react to 5% vinegar but the reactions are more muted. For example, whole calcite reacts strongly to 5% HCl but only weakly to 5% vinegar. Only if this mineral is powdered first will 5% vinegar react strongly with it. In the case of whole dolomite, the reaction for 5% HCl is weak but there is no reaction to 5% vinegar. If, however, the dolomite is powdered, then the reaction is strong with 5% HCl and weak with 5% vinegar. The 'apple cider vinegar' sold in grocery stores, another common food ingredient, is also acetic acid and has a strength similar to white vinegar. The former, however, is brown-colored and often murky whereas the latter is clear and colorless. Although apple cider vinegar can be employed to test minerals, its lack of clarity makes it less useful.

3.2.9 Reaction of Carbonate Minerals to Alizarin Red S Dye

Besides their reaction to acid, carbonate-bearing minerals can also be distinguished by how they are stained with the organic dye known as Alizarin Red S. This red powder is used in histopathological studies to stain calcium deposits in tissues and bones, and is available from the same sources as the reagent-grade hydrochloric acid, i.e. chemical supply companies and university chemistry stockrooms. A staining solution for testing minerals consists of 1 part reagent-grade HCl acid to 499 parts water (preferably distilled or deionized), making 0.2% volume percent hydrochloric acid, plus 0.1 g of dye per 100 ml of solution. Thus, for a 100 ml solution, add 0.2 ml of HCl acid to 99.8 ml of water along with 0.1 g of dye. Mix completely by stirring. Such small amounts of acid and dye may be difficult to measure out in the field and so it will be easier to prepare a larger quantity of staining solution, say 1000 ml (1 liter) with 2 ml of acid added to 998 ml of water and 1 g of dye. In practice, however, the proportions are not critical and so approximate amounts work just as well.

The purpose of the hydrochloric acid is to soften-up, but not dissolve, the mineral surface to make it more receptive to the dye. Although not nearly as dangerous to handle as the concentrated acid, the same precautions should nevertheless be taken when working with Alizarin Red S. This dye comes as a powder that can easily become airborne if breathed upon or exposed to a strong air current. It would be an irritant to skin, eyes, or lungs but toxic only if ingested.

Only a single drop of the staining solution is needed to test a mineral specimen and this is best applied with a dropper. The surface tested, needless to say, should be free of all dirt and oils. Allow 2-3 minutes for the dye to react with the mineral. The following results will be observed: [1] calcite ($CaCO_3$ monoclinic) and aragonite ($CaCO_3$ orthorhombic) are stained red or pink; and [2] ankerite ($CaFe[CO_3]_2$), iron-bearing (impure) dolomite ($CaMg[CO_3]_2$), and cerussite ($PbCO_3$) are stained purple. Pure dolomite does not stain.

3.2.10 *Solubility in Water and Taste*

Some minerals are soluble (i.e. dissolve) in fresh water and so this can be a useful diagnostic property. The dissolution process can be unbearably slow for whole specimens in cold water, but it can be greatly accelerated by powdering the specimen first to increase its reactive surface area, or by using warm or, better, hot (but not boiling) water. Do both to maximize the dissolution rate.

Taste is a diagnostic property for some minerals that are soluble in water and, hence, also in saliva. It can be tested by touching the tip of the tongue to the mineral, but fastidious people may prefer to touch the mineral with a wetted finger and then taste the finger. Mineral tastes have been variously and inconsistently described in the mineralogical literature, often by referencing a substance you are not likely to have tasted before or, in some cases, would not want to taste. It is more useful to refer instead to the primary tastes that the human palette is sensitive to: **salty** (due to the presence of sodium or potassium); **sweet** (due to the presence of an alkali metal salt); **sour** (due to the presence of acid, pH = <7); or **bitter** (due to the presence of alkali, pH = >7; sometimes perceived as metallic). **Astringent** is often included with the mineral tastes but this is really a sensation, not a taste. It is caused by a drying of the tongue and palette that leaves them feeling rough.

3.2.11 *Reaction to a Blowpipe Flame*

3.2.11.1 Blowpipe Analysis

Prior to the widespread use of x-ray diffraction for mineral identification in the mid 1900s, the identification process was routinely accomplished, in combination with other macroscopic observations, by a battery of physical and chemical tests utilizing a high-temperature flame and referred to collectively as 'blowpipe analysis'. These tests involved the use of sometimes toxic or corrosive chemicals to enhance reactions, wood charcoal and gypsum plaster substrates to facilitate the formation and observation of mineral residues and sublimates, and glass tubes that are open either at both ends or just one (i.e. a standard test tube) to foster reactions in oxidizing and reducing environments, respectively. You can find detailed information on these tests in practically any of the early mineralogy books. Those of Brush (1905) and Moses and Parsons (1916) are especially good, and more recent summaries are provided by Hurlbut (1966: 209-220) and Gribble (1988: 445-457). For the purposes of the present book, however, we need only concern ourselves with the simpler and safer blowpipe tests: mineral behavior in an open flame (i.e. fusibility and flame color) and in a heated test tube (i.e. fusibility, condensation, and sublimation).

The flame can come from either a candle, a small wicked-lamp burning alcohol, olive oil, or other liquid fuel, or a small gas (butane or propane) torch. A candle is the safest and easiest of these to use, and is recommended here. The flame is focused, and thereby intensified, by blowing oxygen-rich air into it through a blowpipe (Figure 3.15). This is a narrow tube of heat-resistant metal that, ideally, has an inside diameter of 4-5 mm, with a larger opening at one end for blowing into and a smaller one (the nozzle) with an opening not more than 1 mm across to expel air from. Although the tube can be straight, it is better if it bends between 120° and 90° near its narrow end. This allows the analyst to blow the flame sideways, making it easier to see the reaction of the mineral. It is also helpful to bend the candle wick in the same direction that the flame is being blown. It is now difficult to find ready-made blowpipes for sale but the occasional geological or jewelers supply company will still carry them. You can also make your own blowpipe. This can be any narrow metal tube from a hardware store that can be bent and pinched to form a nozzle. The objective, when working with a blowpipe, is to expel air from the blowpipe in a continuous, uniform blast. Experienced analysts have learned to maintain a steady flow of air by breathing through the nose while simultaneously blowing from the mouth. This is not easily done and requires practice to master. For the mineral testing described below, however, periodic, sustained blasts of air between breaths is good enough.

A candle flame has three basic color zones (Figure 3.15 inset): a dark one at the base of the flame (around and just above the wick) that varies from shades of red, orange and brown and has temperatures of between 800°C and 1000°C; a larger, brightly luminous yellow zone in the upper part of the flame with a temperature of about 1200°C and reducing (oxygen-poor) conditions; and surrounding both, a thin, faintly bluish to white zone extending from the base of the flame, where it is most visible, along the margins of the yellow zone with a temperature of about 1400°C and oxidizing (oxygen-rich) conditions. The hottest part of

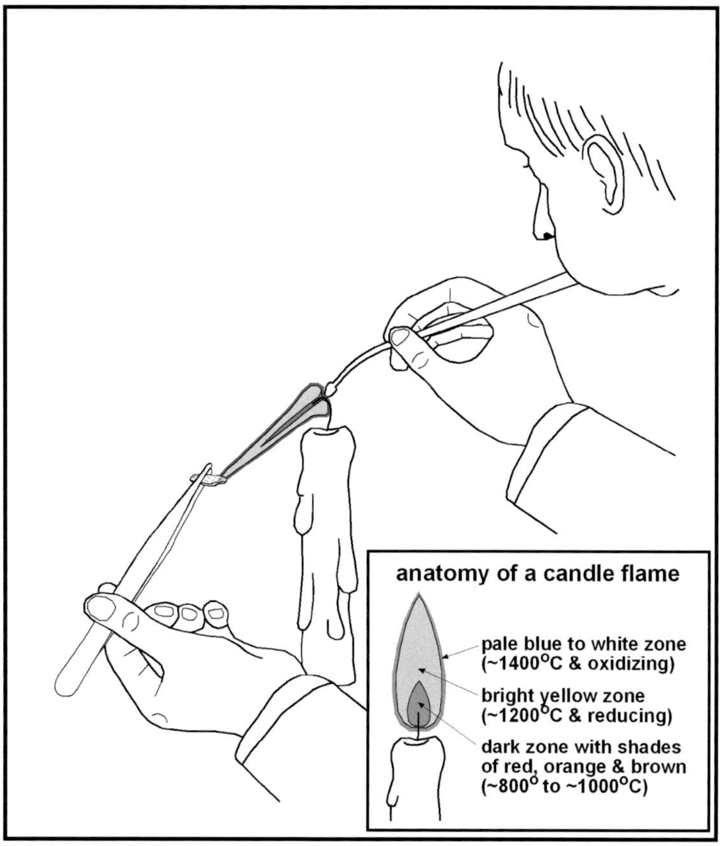

Figure 3.15: Use of a blowpipe with a candle flame.

the flame is near the tip, where the yellow and blue/white zones converge. As reflected in the older mineralogical literature, it was originally thought that the flame was hottest at the tip of the dark zone. The blowpipe nozzle is held near the top of the wick, either just inside the flame or at its edge. The flame will narrow and elongate when blown with a blowpipe, but the color zones remain. The extra oxygen delivered by the blowpipe increases the flame temperatures but by how much is unclear. When using a blowpipe, a mineral fragment is held at the tip of the flame using either steel forceps or tweezers (Figure 3.15) and heated until it becomes 'red hot'. If your hand is steady enough, you can also place the fragment on the tip of a knife blade and then hold this in the flame. If the mineral is powdered, then a knife blade is your only option.

3.2.11.2 Fusibility Test

The **fusibility** of a mineral is the ease with which it can be melted. This depends more on how well a mineral conducts heat than on its melting temperature. Given two minerals with the same melting temperature, the better thermal conductor will be more difficult to melt (i.e. fuse) than the poor conductor which allows the heat to build up more quickly. The test is done by holding the mineral fragment within the hottest part of the blowpipe flame. To facilitate the reaction, the specimen should ideally be either an elongated splinter or a flat chip of no more than about 5 mm across. Such shapes have a large surface area-to-mass ratio and this speeds up the thermal reaction. Table 3.2 provides the fusibility scale employed by mineralogists, but with a 0 fusibility value added here in order to include the few minerals that would otherwise be excluded at the low end of the scale. The fusibility values from this scale are reported for the minerals in Tables 3.4-3.6, and these minerals are also listed in order of increasing fusibility in Table 3.11. Two fusibility ranges are used in the organization of Table 3.4-3.6: ≤ 3 and > 3. Whereas the higher fusibility values are more difficult to discern, a fusibility of 3 or less is easy to recognize because the fusion (or other decomposition response) happens quickly and completely.

Table 3.2: Fusibility scale for minerals.

Fusibility	Observed Reaction
0	quickly volatizes to gas or burns (combusts) without fusing
1	a powder fuses easily in a test tube and a chip fuses easily in an unblown candle flame
2	a powder fuses with difficulty in a test tube, but a chip fuses easily only in a blowpipe flame
3	a chip fuses easily to a globule in a blowpipe flame but the powder not at all in a test tube
4	a chip fuses easily only on its sharp, thin edges (which then become rounded) in a blowpipe flame
5	a chip fuses with difficulty on the sharp, thin edges in a blowpipe flame
6	a chip fuses only on the finest points and thinnest edges in a blowpipe flame
7	infusible in a blowpipe flame

3.2.11.3 Flame Color Test

Some elements can impart a particular color to a blowpipe flame and the intensity of the coloration is proportional to both the amount of the element in a mineral and the degree to which a mineral volatizes in a flame. The color is seen in the flame coming off of (behind) the specimen, and this is facilitated by blowing the flame along the specimen's edge. Because it can be faint, the flame color is best seen in a dark room or against a dark background. The coloration can be either brief or persistent depending on the element and how quickly it is volatized. It can also be enhanced by first wetting the specimen in dilute hydrochloric acid. When multiple color-producing elements are present in a mineral, their flame colors may mask or modify each other. This is the case, for example, with potassium and sodium, which commonly occur in the same minerals. The elements and their characteristic flame colors are provided in Table 3.3. Keep in mind, however, that the description of a flame's color is somewhat subjective and, consequently, you need to allow for some leeway in interpreting these colors.

Table 3.3: Blowpipe flame colors associated with certain elements.

Element	Flame Color
antimony	pale-green, but rarely seen
arsenic	pale-blue
barium	yellowish-green
boron	yellowish-green, but only briefly seen
calcium	slightly orange or orangish-red to mainly yellowish-red
chlorine	slightly purplish-blue
copper, most compounds	vivid dark-green
copper, chloride compounds	slightly purplish-blue
lead	pale-purplish-blue sometimes tinged with green or, rarely, pale-green
lithium	slight to intense purplish-red, fairly persistent
phosphorous	pale-bluish-green to greenish-blue
potassium	pale-violet, fairly persistent
sodium	intense yellow, very persistent
strontium	slightly orangish-red to purplish-red
sulfur	pale-blue
zinc	bluish-green in bright flashes

3.2.11.4 Other Heating Tests

Minerals that release abundant gas during heating (e.g. H_2O or CO_2) will usually experience either **decrepitation** (an explosive and often audible disintegration) or **intumescence** (swelling and bubbling). **Induced magnetism** is another response to heating, which occurs primarily in iron-rich minerals. This can be a strengthening of an originally weak magnetism or the generation of magnetism where there originally was none. In both cases, the high temperature allows the unpaired electrons in the iron atoms to align with the Earth's magnetic field and this effect is then temporarily preserved after the mineral cools.

The so-called 'hot point' (a.k.a. 'hot pin or needle') test is used to identify heat-sensitive materials of biological origin (see Table 5.18). A pointed steel tool (e.g. a pin, needle, or knife point) is heated in a candle or other flame until it is red hot. The point is then pressed into a material and the result observed: does it soften or scorch the material, or cause it to emit an odor? This test, of course, accomplishes then same thing as holding a specimen in a flame, but is much less damaging, as the reaction occurs only at the tool's tip.

Some minerals undergo diagnostic reactions when heated in a Pyrex glass test tube. The bottom of the tube should be held in the hottest part of the candle (or blowpipe) flame with the tube tilted to the side. The following outcomes are observable in the test tube: [1] in the bottom – melting if the specimen is powdered, and decrepitation and intumescence in whole specimens; and [2] on the walls – condensation of water vapor and solid sublimates. The smells coming out of the test tube and the colors of the reaction products can also be clues to a mineral's identity as noted in Tables 3.4-3.6. Yet another use of test tubes is to investigate a mineral's reaction to heated acid or solubility in heated water.

Heating in a test tube is especially useful for those minerals where the fusibility, decrepitation, or intumescence reactions occur too quickly or uncontrollably. It is also an excellent method for distinguishing the two similar-looking calcium sulfate minerals: water-bearing gypsum ($CaSO_4 \cdot 2H_2O$) and water-free anhydrite ($CaSO_4$). Upon heating, gypsum readily loses its water (H_2O) and this dehydration reaction leaves the test tube walls coated with water droplets that condensed from the released water vapor. This reaction will happen with small, whole pieces of gypsum, but occurs quicker and more completely if this mineral is first powdered. Similarly, other water-bearing minerals (e.g. limonite, ~$FeO(OH) \cdot H_2O$) will produce a water condensate when heated in a test tube.

3.3 Determinative Tables

It may surprise you to know that the descriptions of a given mineral species in the principal mineralogical references almost always differ in their details. This is not unexpected given that these descriptions come from different people with different perceptions and experiences and who are working with different mineral specimens. The mineral descriptions in Tables 3.4-3.6, therefore, have been broadened to include all the attributes claimed by the authoritative references consulted (i.e. Sinkankas 1972; Berry et al. 1983; Gribble 1988; Klein and Dutrow 2007; *mindat.org*). In addition, many of the 'other' properties in the rightmost column of the tables were identified or verified by the present author's own investigations. There is more information published on some minerals than on others and this is reflected in their descriptions. Either a 'multi-property' or 'single-property' approach can be taken in keying-out an unknown mineral.

Multi-Property Approach

This employs several macroscopic properties that are arranged in a hierarchy. Navigating through them, from a higher to a lower level, will lead you to an identification. Tables 3.4-3.6 represent this approach and are organized as follows.

- I. Minerals with a metallic or submetallic luster (Table 3.4)
    - A. Mohs hardness ≤ 2.5
        - 1. Streak shades of white, gray, or black
            - i. Fusibility ≤ 3
                - Specific gravity > 2.70
            - ii. Fusibility > 3
                - a. Specific gravity ≤ 2.70
                - b. Specific gravity > 2.70
        - 2. Streak shades of brown, yellow, or red
            - Fusibility ≤ 3 and specific gravity > 2.70
    - B. Mohs hardness > 2.5 to ≤ 5.5
        - 1. Streak shades of white, gray, or black
            - i. Fusibility ≤ 3
                - Specific gravity > 2.70
            - ii. Fusibility > 3 or not determined
                - Specific gravity > 2.70
        - 2. Streak shades of brown, yellow, or red
            - i. Fusibility ≤ 3
                - Specific gravity > 2.70
            - ii. Fusibility > 3
                - Specific gravity > 2.70
    - C. Mohs hardness > 5.5
        - 1. Streak shades of white, gray, or black
            - i. Fusibility ≤ 3
                - Specific gravity > 2.70
            - ii. Fusibility > 3
                - Specific gravity > 2.70
        - 2. Streak shades of brown, yellow, or red
            - i. Fusibility ≤ 3
                - Specific gravity > 2.70
            - ii. Fusibility > 3
                - Specific gravity > 2.70
- II. Minerals with a nonmetallic luster and a colored (non-white) streak (Table 3.5)
    - A. Mohs hardness ≤ 2.5
        - 1. Streak shades of brown, yellow, or red
            - i. Fusibility ≤ 3

a. Specific gravity ≤ 2.70
b. Specific gravity > 2.70
ii. Fusibility not determined
Specific gravity > 2.70
2. Streak shades of blue or green
i. Fusibility ≤ 3
a. Specific gravity ≤ 2.70
b. Specific gravity > 2.70
ii. Fusibility > 3
a. Specific gravity ≤ 2.70
b. Specific gravity > 2.70
B. Mohs hardness > 2.5 to < 5.5
1. Streak shades of brown, yellow, or red
Fusibility > 3 and specific gravity > 2.70
2. Streak shades of blue or green
i. Fusibility ≤ 3
Specific gravity > 2.70
ii. Fusibility > 3 or not determined
a. Specific gravity ≤ 2.70
b. Specific gravity > 2.70
C. Mohs hardness > 5.5
1. Streak shades of brown, yellow, or red
Fusibility > 3 and specific gravity > 2.70
2. Streak shades of gray or black
Fusibility > 3 and specific gravity > 2.70
III. Minerals with a nonmetallic luster and either a white or white grading to a pale-colored streak (Table 3.6)
A. Mohs hardness ≤ 2.5
1. Fusibility ≤ 3
a. Specific gravity ≤ 2.70
b. Specific gravity > 2.70
2. Fusibility > 3 or not determined
a. Specific gravity ≤ 2.70
b. Specific gravity > 2.70
B. Mohs hardness > 2.5 to ≤ 5.5
1. Fusibility ≤ 3
Specific gravity > 2.70
2. Fusibility > 3
a. Specific gravity ≤ 2.70
b. Specific gravity > 2.70
C. Mohs hardness > 5.5
1. Fusibility ≤ 3
Specific gravity > 2.70
2. Fusibility > 3 or not determined
a. Specific gravity ≤ 2.70
b. Specific gravity > 2.70

Within each of the above divisions, minerals are listed downward in order of increasing minimum hardness and, when they have the same hardness, downward in order of increasing minimum specific gravity. Some minerals appear in more than one division within a given table or even in two different tables when their properties range across the arbitrary division boundaries. The following sequential steps are recommended to identify an unknown mineral specimen using Tables 3.4-3.6.

1. Observe whether the specimen has a metallic, submetallic, or nonmetallic luster.
2. Determine the color of the specimen's streak or powdered form.
3. Observe the dry color of the whole mineral.
4. Determine in which Mohs hardness range the specimen falls: ≤ 2.5, > 2.5 to ≤ 5.5, or > 5.5. If you can measure the hardness more precisely then, of course, this will be very useful.
5. If you can obtain a small chip or splinter from the specimen, determine in which fusibility range it falls: ≤ 3 or > 3.
6. If the mineral has not been identified by this point, the number of possible candidates will, at least, be very small. To further reduce the possibilities, look at the 'other' properties described for each mineral in the rightmost column of the tables. Begin with cleavage and habit, and continue, if necessary, with the other properties.

If your specimen does not seem to match any mineral in Tables 3.4-3.6, then you probably have either a very impure specimen, a rare form of one of the tabled minerals, or a mineral not in these tables. In the latter case, the mineral will almost certainly be found in Table 3.9.

Single-Property Approach

The idea here is to employ one property to quickly arrive at a (hopefully) short list of minerals that includes your unknown. Specific gravity is the single most useful property for this approach. The mineral values are tabulated in Table 3.9 in order of increasing specific gravity, and include not only the 150 minerals in Tables 3.4-3.6, but also another 200 minerals that are widely considered to be geologically common or economically important. These latter minerals need to be included because it is entirely possible (and even likely) that you will eventually encounter some of them in your excavation or survey. Once you have a short list of possible minerals, use the book's index to find the entry for each mineral on the list in Tables 3.4-3.6 and see if its properties match those of your unknown. Two other measurements can also be used in the single-property approach: Mohs hardness in Table 3.10 and fusibility in Table 3.11. These tables, however, are limited to just the minerals in Tables 3.4-3.6.

Table 3.4: Determination of archaeologically significant and common rock-forming minerals with a metallic or submetallic luster.

I. LUSTER = METALLIC OR SUBMETALLIC				
A. Mohs Hardness ≤ 2.5				
1. Streak shades of white, gray, or black				
i. Fusibility ≤ 3				
Specific Gravity > 2.70				
Streak Color	**Mineral Color**	**Specific Gravity**	**Mohs Hardness**	**Mineral Name and Other Properties**
no streak but a pale-gray liquid smear	pale-gray	13.59	liquid	**native mercury, Hg** [a.k.a. quicksilver] *(technically neither a mineral nor mineraloid)* • breakage & tenacity: liquid • habit: always in small fluid globules • luster: metallic • diaphaneity: opaque • fusibility: 0 (volatizes) □ use: mixed with sulfur to make synthetic vermillion, a bright red paint pigment
gray to black	dark-purplish-blue (tarnishing to iridescent red and yellow)	4.60-4.76	1.5-2	**covellite, CuS** [a.k.a. covelline and blue or indigo copper] *(hexagonal)* • breakage: cleavage in 1 direction (basal pinacoidal), prominent • tenacity: flexible • habit: usually in aggregates that are internally granular and externally massive; and occasionally in hexagonal tabular crystals • luster: metallic • diaphaneity: pseudo-opaque to opaque • fusibility: 2-2.5 • other: small pieces easily ignite and burn in a blowpipe flame while emitting sulfurous fumes (the burning is seen as a thin blue haze surrounding the piece and is best observed in a dark room immediately after removal from the blowpipe flame); yields a yellow sulfur sublimate when heated in a test tube □ use: smelted for copper

gray to black	gray (tarnishing to iridescent black)	4.52-4.63	2	**stibnite, Sb_2S_3** [a.k.a. antimonite, antimony glance, and gray antimony] *(orthorhombic)* • breakage: cleavage in 4 directions (prismatic + two pinacoidal), none prominent; also sub-conchoidal fracture • tenacity: slightly flexible and sectile • habit: usually in short acicular to bladed prismatic crystals, often curved or bent, with longitudinally striated faces and commonly in radiated groups of crystals; and occasionally in aggregates that are internally fine to coarse granular and externally massive or stalactitic • luster: metallic • diaphaneity: opaque • fusibility: 1 • other: gives off sulfurous fumes and yields a white sublimate when heated in a test tube □ use: possibly gray paint pigment; possibly smelted for antimony
shiny dark-gray to black	shiny dark-gray to black (tarnishing to dull black)	7.19-7.36	2-2.5	**acanthite, Ag_2S** [a.k.a. argentite and silver glance] *(isometric or monoclinic depending on crystallization conditions)* • breakage: cleavage in 9 directions (cubic + dodecahedral), none prominent; also uneven to sub-conchoidal fracture • tenacity: sectile • habit: usually in aggregates that are internally fine granular and externally massive or dendritic; and rarely in cubical, octahedral, dodecahedral, or filiform crystals • luster: metallic • diaphaneity: opaque • fusibility: 1.5 • other: fuses with intumescence in a blowpipe flame (a hand-lens may be needed to see the bubbled surface); gives off sulfurous fumes when heated in a test tube □ use: possibly smelted for silver

gray	gray	7.40-7.58	2.5	**galena, PbS** [a.k.a. galenite, lead glance, and blue lead] (isometric) • breakage: cleavage in 3 directions (cubic), all prominent; also even to sub-conchoidal fracture • tenacity: brittle • habit: commonly in cubical or octahedral crystals; and also commonly in aggregates that are internally fine to coarse granular and externally massive • luster: metallic • diaphaneity: opaque • fusibility: 1 • other: when heated in a test tube fuses with intumescence to metallic lead and gives off sulfurous fumes; colors a blowpipe flame bluish-green but this can be difficult to see □ use: gray paint pigment; colorant in vitreous glazes; jewelry and small decorative objects; smelted for lead and occasionally silver, a sometimes abundant impurity; gold refining
A. Mohs Hardness ≤ 2.5				
1. Streak shades of white, gray, or black				
ii. Fusibility > 3				
a. Specific Gravity ≤ 2.70				
shiny black	dark-gray to black	2.00-2.23	1-2	**graphite, C** [a.k.a. native carbon, plumbago, and black lead] (hexagonal) • breakage: cleavage in 1 direction (basal pinacoidal), prominent with easy separation but difficult to see • tenacity: flexible and sectile • habit: usually in aggregates that are internally granular to mainly foliated and externally massive or stalactitic; and rarely in hexagonal tabular crystals occurring in radiated groups • luster: metallic to submetallic • diaphaneity: opaque • fusibility: 7 • other: greasy feel □ use: black ink and paint pigment; small carved objects

<table>
<tr><td colspan="5">A. Mohs Hardness ≤ 2.5</td></tr>
<tr><td colspan="5">1. Streak shades of white, gray, or black</td></tr>
<tr><td colspan="5">ii. Fusibility > 3</td></tr>
<tr><td colspan="5">b. Specific Gravity > 2.70</td></tr>
<tr><td>white to pale-gray</td><td>usually dark-brown or black (weathers to yellow); rarely dark-green for biotite
—
usually pale-yellow or yellowish-brown; rarely greenish- to reddish-brown for phlogopite</td><td>2.70-3.30</td><td>2-3</td><td>biotite, $K(Mg,Fe)_3(AlSi_3O_{10})(OH)_2$ to phlogopite, $KMg_3(AlSi_3O_{10})(OH)_2$ in a solid-solution series [mica group; a.k.a. black mica for biotite; and amber, brown or bronze mica for phologopite] (monoclinic)
• breakage: cleavage in 1 direction (basal pinacoidal), prominent with easy separation
• tenacity: flexible and elastic
• habit: commonly in pseudo-hexagonal tabular or short prismatic crystals; and also commonly in aggregates that are internally foliated and externally massive
• luster: submetallic to mainly nonmetallic vitreous or pearly
• diaphaneity: transparent to translucent
• fusibility: 5
• other: yields water condensate when heated in a test tube; immediately whitens in a blowpipe flame with fusion along the edges; biotite is weakly attracted to a neodymium magnet
□ use: none known, but these are common rock-forming minerals</td></tr>
<tr><td>black to bluish-black</td><td>light- to dark-gray or black, commonly bluish</td><td>4.40-4.75 (aggregates)

up to 5.06 (crystals)</td><td>2-6 (aggregates)

6-6.5 (crystals)</td><td>pyrolusite, MnO_2 [a.k.a. polianite] (tetragonal)
• breakage: no cleavage; uneven to hackly or splintery fracture
• tenacity: brittle
• habit: usually in aggregates that are internally finely granular or fibrous and externally massive, radiated, stalactitic, or dendritic; and rarely in prismatic crystals (commonly pseudomorphous after other mineral crystals)
• luster: metallic to submetallic
• diaphaneity: opaque
• fusibility: 7
• other: turns brown in a blowpipe flame but sometimes only in patches; yields a little water when heated in a test tube; sometimes weakly attracted to a neodymium magnet
□ use: black paint pigment</td></tr>
</table>

A. Mohs Hardness ≤ 2.5				
2. Streak shades of brown, yellow, or red				
Fusibility ≤ 3 and Specific Gravity > 2.70				
bright red	brownish-red when impure to bright red when pure	8.09-8.10	2-2.5	**cinnabar, HgS** [a.k.a. vermillion] *(hexagonal)* • breakage: cleavage in 3 directions (rhombohedral), all prominent; also uneven to sub-conchoidal fracture • tenacity: slightly sectile • habit: commonly in hexagonal prismatic to tabular and rhombohedral crystals; and also commonly in aggregates that are internally fine granular and externally massive • luster: submetallic to mainly nonmetallic adamantine (crystals) or dull (aggregates) • diaphaneity: transparent to translucent • fusibility: 0 (volatizes) • other: volatizes when heated in a test tube, giving off toxic mercuric fumes and yielding a black sublimate of mercuric sulfide that becomes red when rubbed after cooling □ use: red paint pigment; jewelry and small decorative objects; gilding when Hg is mixed with gold and silver; smelted for mercury; medicine

<table>
<tr><td colspan="5">B. Mohs Hardness > 2.5 to ≤ 5.5</td></tr>
<tr><td colspan="5">1. Streak shades of white, gray, or black</td></tr>
<tr><td colspan="5">i. Fusibility ≤ 3</td></tr>
<tr><td colspan="5">Specific Gravity > 2.70</td></tr>
<tr><td>white to pale-gray</td><td>usually dark-brown or black (weathers to yellow); rarely dark-green for biotite
—
usually pale-yellow or yellowish-brown; rarely greenish- to reddish-brown for phlogopite</td><td>2.70-3.30</td><td>2-3</td><td>biotite, $K(Mg,Fe)_3(AlSi_3O_{10})(OH)_2$ to phlogopite, $KMg_3(AlSi_3O_{10})(OH)_2$ in a solid-solution series [mica group; a.k.a. black mica for biotite; and amber, brown or bronze mica for phologopite] (monoclinic)
• breakage: cleavage in 1 direction (basal pinacoidal), prominent with easy separation
• tenacity: flexible and elastic
• habit: commonly in pseudo-hexagonal tabular or short prismatic crystals; and also commonly in aggregates that are internally foliated and externally massive
• luster: submetallic to mainly nonmetallic vitreous or pearly
• diaphaneity: transparent to translucent
• fusibility: 5
• other: yields water condensate when heated in a test tube; immediately whitens in a blowpipe flame with fusion along the edges; biotite is weakly attracted to a neodymium magnet
□ use: none known, but these are common rock-forming minerals</td></tr>
<tr><td>gray to black</td><td>gray to black</td><td>5.83</td><td>2.5-3</td><td>bournonite, $PbCuSbS_3$ [a.k.a. berthonite, endellionite, cogwheel ore, and antimonial copper glance] (orthorhombic)
• breakage: cleavage in 3 directions (all pinacoidal), none prominent; also sub-conchoidal fracture
• tenacity: brittle
• habit: usually in aggregates that are internally granular and externally massive; and occasionally in short prismatic or tabular crystals with alternating shiny and dull faces that are marked by parallel striations
• luster: metallic to submetallic
• diaphaneity: opaque
• fusibility: 1
• other: colors a blowpipe flame green but this can be difficult to see
□ use: smelted for copper</td></tr>
</table>

shiny dark-gray	shiny dark-gray (tarnishing to dull black)	5.50-5.80	2.5-3	**chalcocite, Cu_2S** [a.k.a. redruthite, chalcosine, vitreous copper, and copper glance] *(monoclinic or orthorhombic depending on crystallization conditions)* • breakage: cleavage in 1 direction (pinacoidal), not prominent; also conchoidal fracture • tenacity: brittle to slightly sectile • habit: usually in aggregates that are internally compact granular and externally massive; and rarely in hexagonal prismatic and tabular crystals, which can occur in stellate groups • luster: metallic • diaphaneity: opaque • fusibility: 2-2.5 • other: colors a blowpipe flame dark-green but this can be difficult to see □ use: smelted for copper
shiny white	light-gray to white (tarnishing to dark-gray, black, or brown)	10.50 pure 10-12 impure	2.5-3	**native silver, Ag** *(isometric)* • breakage: no cleavage; hackly fracture • tenacity: ductile, malleable, and sectile • habit: commonly in filiform crystals and rarely in distorted cubical, octahedral, or dodecahedral crystals; and also commonly in aggregates that are internally granular and externally dendritic • luster: metallic • diaphaneity: opaque • fusibility: 2 • note: **electrum** is a mixture of native gold and silver □ use: jewelry and decorative arts; tableware; mirrors; coinage
black	dark-gray to black	4.40-4.50	3	**enargite, Cu_3AsS_4** *(orthorhombic)* • breakage: cleavage in 4 directions (prismatic + two pinacoidal), only two (prismatic) prominent; also uneven fracture • tenacity: brittle • habit: commonly in tabular to long bladed prismatic crystals with longitudinal striations; and also commonly in aggregates that are internally granular and externally massive • luster: metallic • diaphaneity: opaque • fusibility: 1 • other: colors a blowpipe flame green but this can be difficult to see □ use: possibly smelted for copper and/or arsenic

dark-gray	red to yellowish-brown (tarnishing to iridescent blue and purple, and occasionally black)	4.90-5.40	3	**bornite, Cu_5FeS_4** [a.k.a. erubescite , purple copper, and peacock ore] *(isometric or tetragonal depending on crystallization conditions)* • breakage: no cleavage; uneven to conchoidal fracture • tenacity: brittle • habit: usually in aggregates that are internally granular and externally massive; and rarely in cubical, octahedral, or dodecahedral crystals • luster: metallic • diaphaneity: opaque • fusibility: 2.5 • other: blackens in a blowpipe flame and becomes dark-reddish-brown upon cooling; colors a blowpipe flame green but this can be difficult to see; weakly attracted to a neodymium magnet but becomes moderately to strongly attracted after heating in a candle or blowpipe flame □ use: smelted for copper
greenish-black	light-brownish-yellow, occasionally with a greenish tinge	5.30-5.70	3-3.5	**millerite, NiS** [a.k.a. capillary, hair or nickel pyrites] *(hexagonal)* • breakage: cleavage in 3 directions (rhombohedral), only two prominent • tenacity: brittle but elastic when filiform • habit: usually in acicular to filiform crystals, commonly in radiated groups; and rarely in aggregates that are internally granular and externally stalactitic • luster: metallic • diaphaneity: opaque • fusibility: 1.5-2 • other: gives off sulfurous fumes when heated in a test tube □ use: possibly smelted for metallic nickel; possibly a greenish colorant for vitreous glazes
dark-gray	white to gray (tarnishing to iridescent yellow and brown, and sometimes developing a dark-brown surface coating)	7.20-8.10	3-3.5	**domeykite, Cu_3As** [a.k.a. arsenical copper pyrite] *(isometric)* • breakage: no cleavage; uneven fracture • tenacity: brittle to slightly sectile • habit: almost always in aggregates that are internally granular and externally massive, botryoidal, or reniform • luster: metallic • diaphaneity: opaque • fusibility: 2 □ use: possibly smelted for arsenical copper

black	gray to black	6.25-6.45	3-4	**tenorite, CuO** [a.k.a. melaconite and black copper] *(monoclinic)* • breakage: no cleavage; uneven to conchoidal fracture • tenacity: brittle to flexible and elastic, and occasionally pulverulent • habit: commonly in acicular to bladed prismatic crystals that are curved and occur in stellate groups; and also commonly in aggregates that are internally granular or foliated and externally massive or botryoidal • luster: metallic • diaphaneity: opaque • fusibility: 3 • other: colors a blowpipe flame dark-green but this can be difficult to see □ use: smelted for copper
brown or black	dark-gray or black	4.50-5.10	3-4.5	**tennantite, $Cu_{12}As_4S_{13}$** to **tetrahedrite, $Cu_{12}Sb_4S_{13}$** in a solid-solution series [a.k.a. panabase and stylotypite for tetrahedrite; and fahlore and gray copper for both] *(isometric)* • breakage: no cleavage; uneven or hackly fracture • tenacity: brittle • habit: usually in aggregates that are internally compact to coarse granular and externally massive; and rarely in tetrahedral crystals • luster: metallic • diaphaneity: pseudo-opaque to mainly opaque • fusibility: 1.5 • other: colors a blowpipe flame green but this can be difficult to see □ use: possibly smelted for copper and silver, a common impurity

greenish-black	yellow (tarnishing to iridescent brownish-yellow)	4.10-4.30	3.5-4.5	**chalcopyrite, $CuFeS_2$** [a.k.a. copper pyrite and yellow copper] *(isometric or tetragonal depending on crystallization conditions)* • breakage: no cleavage; uneven to conchoidal fracture • tenacity: brittle to slightly sectile • habit: commonly in wedge-shaped tetrahedral crystals; and also commonly in aggregates that are internally fine granular and externally massive or botryoidal • luster: metallic • diaphaneity: opaque • fusibility: 2-2.5 • other: sometimes colors a blowpipe flame dark-green but this is seldom visible; when heated in a test tube or blowpipe flame, decrepitates and becomes colorfully iridescent; sometimes weakly attracted to a neodymium magnet □ use: smelted for copper
dark-gray to black	yellowish-brown to occasionally reddish-brown (tarnishing to darker shades)	4.58-4.65	3.5-4.5	**pyrrhotite, $Fe_{1-x}S$** (where x = 0 to 0.2) [a.k.a. dipyrite, pyrrhotine, and magnetic pyrite] *(monoclinic or hexagonal depending on crystallization conditions)* • breakage: no cleavage; uneven to sub-conchoidal fracture • tenacity: brittle • habit: usually in aggregates that are internally granular and externally massive; and rarely in hexagonal tabular crystals • luster: metallic • diaphaneity: opaque • fusibility: 2.5-3 • other: strongly attracted to a neodymium magnet □ use: none known, but it is a common rock-forming mineral
black	dark-gray when pure, otherwise black or yellowish-brown (tarnishing to pale-blue)	4.4	4	**stannite, Cu_2FeSnS_4** [a.k.a. bell-metal ore, stannine, and tin pyrite] *(tetragonal)* • breakage: cleavage in 3 directions (prismatic + basal pinacoidal), none prominent; also uneven fracture • tenacity: brittle • habit: almost always in aggregates that are internally granular and externally massive • luster: metallic • diaphaneity: opaque • fusibility: 1.5 • other: decrepitates in a candle flame; gives off sulfurous fumes when heated in a test tube and yields a whitish tin oxide sublimate □ use: smelted for copper and tin

B. Mohs Hardness > 2.5 to ≤ 5.5				
1. Streak shades of white, gray, or black				
ii. Fusibility > 3 or not determined				
Specific Gravity > 2.70				
black to bluish-black	light- to dark-gray or black, commonly bluish	4.40-4.75 (aggregates) up to 5.06 (crystals)	2-6 (aggregates) 6-6.5 (crystals)	**pyrolusite**, **MnO_2** [a.k.a. polianite] *(tetragonal)* • breakage: no cleavage; uneven, hackly or splintery fracture • tenacity: brittle • habit: usually in aggregates that are internally finely granular or fibrous and externally massive, radiated, stalactitic, or dendritic; and rarely in prismatic crystals (commonly pseudomorphous after other mineral crystals) • luster: metallic to submetallic • diaphaneity: opaque • fusibility: 7 • other: turns brown in a blowpipe flame but sometimes only in patches; yields a little water when heated in a test tube; sometimes weakly attracted to a neodymium magnet □ use: black paint pigment
dark-gray	dark-gray (tarnishing to bluish-black)	5.55-5.71	2.5-3	**digenite**, **Cu_9S_5** [a.k.a. blue chalcocite and carmenite] *(hexagonal)* • breakage: no cleavage; conchoidal fracture • tenacity: brittle • habit: almost always in pseudo-cubical crystals • luster: metallic • diaphaneity: opaque • fusibility: 3 or less □ use: possibly smelted for copper
shiny gray	light- to dark-gray	21.46 pure 14-19 impure	4-4.5	**native platinum**, **Pt** [a.k.a. white gold] *(isometric)* • breakage: no cleavage; hackly fracture • tenacity: ductile and malleable • habit: usually in irregular grains; and occasionally in aggregates that are internally granular and externally dendritic • luster: metallic • diaphaneity: opaque • fusibility: 7 □ use: an additive for hardening gold

black	light- to dark-gray or black	4.93-4.95	4-6	**hollandite, $BaMn_8O_{16}$** *(monoclinic or tetragonal depending on crystallization conditions)* • breakage: cleavage in 2 directions (prismatic), none prominent • tenacity: brittle • habit: commonly in short, often acicular, prismatic crystals; and also commonly in aggregates that are internally fibrous and externally massive or colloform • luster: metallic • diaphaneity: opaque • fusibility: not determined □ use: black paint pigment
gray to black	gray to black	7.30-7.90	4.5	**meteoritic iron, Fe** *(isometric)* • breakage: no cleavage; hackly fracture • tenacity: malleable • habit: always in aggregates that are internally granular to foliated and externally massive • luster: metallic • diaphaneity: opaque • fusibility: 7 • other: strongly attracted to neodymium and other weaker magnets □ use: jewelry; tools and weapons
shiny brownish-black to black	dark-gray to black	3.70-4.70	5-6	**romanèchite, $(Ba,H_2O)_2(Mn^{4+},Mn^{3+})_5O_{10}$** [a.k.a. black hematite and psilomelane] (orthorhombic) – converts to hollandite at high temperatures • breakage: no cleavage; uneven fracture • tenacity: brittle • habit: always in aggregates that are internally granular and externally massive, colloform, or stalactitic • luster: submetallic • diaphaneity: opaque • fusibility: 7 • other: yields some water condensate when heated in a test tube □ use: black paint pigment

B. Mohs Hardness > 2.5 to ≤ 5.5				
2. Streak shades of brown, yellow, or red				
i. Fusibility ≤ 3				
Specific Gravity > 2.70				
shiny red	light-red (tarnishing to dark-red or brown, and commonly coated with green Cu-oxide corrosion)	8.94 pure	2.5-3	**native copper**, **Cu** *(isometric)* • breakage: no cleavage; hackly fracture • tenacity: malleable and ductile • habit: commonly in cubical, octahedral, dodecahedral, or filiform crystals; and also commonly in aggregates that are internally granular and externally dendritic • luster: metallic • diaphaneity: opaque • fusibility: 3 • other: colors a blowpipe flame a conspicuous bright green □ use: jewelry and decorative objects; tools and weapons (including those of arsenical and tin bronzes); tableware; mirrors; coinage
shiny yellow in various shades depending on impurities	bright yellow; becomes paler or redder with silver or copper impurities, respectively	19.3 pure 12-20 impure	2.5-3	**native gold**, **Au** *(isometric)* • breakage: no cleavage; hackly fracture • tenacity: ductile, malleable, and sectile • habit: usually in irregular grains; commonly in filiform, but rarely cubical, octahedral, or dodecahedral crystals; and also commonly in aggregates that are internally granular and externally dendritic • luster: metallic • diaphaneity: opaque • fusibility: 2.5 • note: **electrum** is a mixture of native gold and silver □ use: jewelry and decorative arts; tableware; coinage
shiny brownish-red	dark-red to reddish-black	6.10-6.15	3.5-4	**cuprite**, **Cu_2O** [a.k.a. ruby copper] *(isometric)* • breakage: cleavage in 3 directions (cubic), none prominent; also uneven to conchoidal fracture • tenacity: brittle • habit: usually in cubical or octahedral crystals and less commonly in dodecahedral or filiform crystals; and occasionally in aggregates that are internally granular and externally massive • luster: submetallic to nonmetallic adamantine (crystals) or dull (aggregates) • diaphaneity: pseudo-opaque • fusibility: 3 • other: colors a blowpipe flame dark-green but this can be difficult to see □ use: smelted for copper

B. Mohs Hardness > 2.5 to ≤ 5.5				
2. Streak shades of brown, yellow, or red				
ii. Fusibility > 3				
Specific Gravity > 2.70				
reddish-brown to light-yellow; rarely white when pure	commonly yellow, yellowish-brown, dark-brown or black; occasionally red, green, or, when pure, white or colorless	3.90-4.10	3.5-4	**sphalerite, ZnS** [a.k.a. blende, zinc blende, mock lead, and black or steel jack] *(isometric)* • breakage: cleavage in 6 directions (dodecahedral), all prominent; also conchoidal fracture • tenacity: brittle • habit: usually in poorly formed, tetragonal or dodecahedral crystals; and occasionally in aggregates that are internally fibrous or fine to coarse granular and externally botryoidal • luster: submetallic to nonmetallic adamantine or resinous • diaphaneity: translucent to transparent • fusibility: 7 • other: gives off a sulfurous smell when crushed; when powdered, emits a rotten-egg smell (from H_2S) in cold dilute HCl acid; commonly fluoresces orange in long-wavelength UV light; pyroelectric and commonly triboluminescent □ use: smelted for zinc, an ingredient of brass
yellow to yellowish-brown	usually yellow to yellowish-brown; sometimes other shades of brown	2.70-4.30	4-5.5	**limonite, ~FeO(OH)•H_2O** [a.k.a. brown and yellow ochers, brown hematite or iron, and bog iron] *(mineraloid or possibly an intergrowth of goethite and lepidocrocite)* • breakage: no cleavage; uneven fracture • tenacity: friable to brittle • habit: always in soft (earthy or ocherous) to hard aggregates that are internally fibrous and externally massive, mammillary, or stalactitic • luster: submetallic to nonmetallic dull or silky • diaphaneity: translucent to opaque • fusibility: 5-5.5 • other: yields water condensate when heated in test tube, dehydrating to hematite; weakly to moderately attracted to a neodymium magnet; commonly has clay mineral impurities when ocherous □ use: smelted for iron; yellow to brownish-yellow paint pigments and colorants in glass and vitreous glazes

brownish- to orangish-yellow or yellowish-brown	dark-brown, reddish- to yellowish-brown, brownish-yellow, and brownish-black	3.30-4.37	5-5.5	**goethite, αFeO(OH)** [a.k.a. allcharite, bog iron, and xanthosiderite] *(orthorhombic)* • breakage: cleavage in 2 directions (both pinacoidal), one prominent; also uneven fracture • tenacity: brittle to pulverulent • habit: usually in aggregates that are internally fibrous to compact granular and externally massive, botryoidal, reniform, oolitic/pisolitic, or stalactitic; and rarely in acicular prismatic, tabular, or filiform crystals (commonly pseudomorphous after cubical pyrite crystals) • luster: submetallic to nonmetallic dull or silky (aggregates), or adamantine (crystals) • diaphaneity: pseudo-opaque to opaque • fusibility: 5.5 • other: yields water condensate when heated in test tube, dehydrating to hematite; weakly to moderately attracted to a neodymium magnet □ use: yellow to brownish-yellow paint pigments and colorants in glass and vitreous glazes; small decorative objects; smelted for iron
red to reddish-brown	gray to rarely black	4.90-5.26	5-6.5	**hematite** (or **haematite**), **Fe_2O_3** *(hexagonal)* metallic variety: **specular hematite** or **specularite** • breakage: no cleavage; uneven to sub-conchoidal fracture • tenacity: brittle to slightly elastic • habit: usually in aggregates that are internally foliated and externally massive; and rarely in hexagonal tabular crystals • luster: metallic • diaphaneity: opaque • fusibility: 7 • other: moderately to strongly attracted to a neodymium magnet □ use: smelted for iron; red paint pigment and colorant for glass and vitreous glazes; small decorative objects; weights; medicine
				submetallic to nonmetallic variety: **common hematite** or **red ocher** [a.k.a. bloodstone, red iron, and kidney ore] • breakage: no cleavage; uneven fracture • tenacity: brittle to friable or pulverulent • habit: usually in aggregates that are internally fibrous or granular and externally massive, colloform, oolitic, pisolitic, or stalactitic

	red to brownish-red			• diaphaneity: pseudo-opaque to opaque • luster: submetallic to nonmetallic dull or silky • fusibility: 7 • other: weakly to moderately attracted to a neodymium magnet; commonly has clay mineral impurities when ocherous □ use: smelted for iron; red paint pigment and colorant for glass and vitreous glazes; small decorative objects; weights; medicine
C. Mohs Hardness > 5.5				
1. Streak shades of white, gray, or black				
i. Fusibility ≤ 3				
Specific Gravity > 2.70				
dark-gray to black	light-gray to white (tarnishing to a faint yellowish tint)	5.90-6.20	5.5-6	**arsenopyrite, $FeAsS$** [a.k.a. arsenical pyrite, mispickel, and white pyrite or mundic] *(monoclinic)* • breakage: cleavage in 1 direction (pinacoidal), not prominent; also uneven fracture • tenacity: brittle • habit: usually in tabular or prismatic, and rarely pseudo-octahedral, crystals with faces marked by parallel striations; and occasionally in aggregates that are internally compact granular and externally massive • luster: metallic • diaphaneity: opaque • fusibility: 2 • other: when heated in a test tube yields a residue of shiny metallic arsenic and a reddish to orangish sublimate of arsenic sulfide that turns yellow after cooling; gives off a garlic-like odor (from arsenic) when scratched, struck, or ground; gives off sparks when struck with quartz or other hard material; weakly attracted to a neodymium magnet □ use: possibly smelted for arsenic or added directly to copper to make arsenical bronze

greenish- to brownish black	pale-yellow to brownish-yellow (tarnishing to darker, iridescent yellow)	4.80-5.02	6-6.5	**pyrite, FeS_2** [a.k.a. iron pyrite, mundic, and fool's gold] *(isometric)* • breakage: no cleavage; uneven to conchoidal fracture • tenacity: brittle • habit: usually in cubical crystals with parallel striations on faces that are perpendicular to those on adjacent faces, and less commonly in octahedral, dodecahedral, or pyritohedral crystals; and occasionally in aggregates that are internally granular and externally massive, reniform, or stalactitic • luster: metallic • diaphaneity: opaque • fusibility: 2.5 • other: ignites and burns in a blowpipe flame (visible as a thin blue haze surrounding the specimen), gives off pungent sulfurous (SO_2) fumes, and leaves a reddish-brown burnt residue; emits a cascade of sparks when struck with quartz or other hard material; although iron-rich, it is not magnetic □ use: strike-stone for starting fires; small mirrors; small decorative objects and inlays
dark-gray to brownish-black	pale-brownish-yellow to almost white on freshly broken surfaces (tarnishing to shades of yellow or brown and occasionally green)	4.89	6-6.5	**marcasite, FeS_2** [a.k.a. cockscomb, spear or white iron pyrites] *(orthorhombic)* • breakage: cleavage in 2 directions (prismatic), none prominent; also uneven fracture • tenacity: brittle • habit: usually in pseudo-hexagonal prismatic or tabular crystals, commonly with cockscomb serration along their edges and in stellate groups; and occasionally in aggregates that are internally fibrous and externally colloform, reniform, or stalactitic • luster: metallic • diaphaneity: opaque • fusibility: 2.5 • other: ignites and burns in a blowpipe flame (visible as a thin blue haze surrounding the specimen), gives off pungent sulfurous (SO_2) fumes, and leaves a reddish-brown burnt residue; emits a sulfurous odor when crushed; yields some yellow sulfur sublimate when heated in a test tube; emits sparks when struck with quartz or other hard material; although iron-rich, it is not magnetic □ use: small decorative objects; strike-stone for starting fires

C. Mohs Hardness > 5.5				
1. Streak shades of white, gray, or black				
ii. Fusibility > 3				
Specific Gravity > 2.70				
black to bluish-black	light- to dark-gray or black, commonly bluish	4.40-4.75 (aggregates) up to 5.06 (crystals)	2-6 (aggregates) 6-6.5 (crystals)	**pyrolusite, MnO_2** [a.k.a. polianite] *(tetragonal)* • breakage: no cleavage; uneven, hackly or splintery fracture • tenacity: brittle • habit: usually in aggregates that are internally finely granular or fibrous and externally massive, radiated, stalactitic, or dendritic; and rarely in prismatic crystals (commonly pseudomorphs after other mineral crystals) • luster: metallic to submetallic • diaphaneity: opaque • fusibility: 7 • other: turns brown in a blowpipe flame but sometimes only in patches; yields a little water when heated in a test tube; sometimes weakly attracted to a neodymium magnet □ use: black paint pigment
shiny brownish-black to black	dark-gray to black	3.70-4.70	5-6	**romanèchite, $(Ba,H_2O)_2(Mn^{4+},Mn^{3+})_5O_{10}$** [a.k.a. black hematite and psilomelane] *(orthorhombic)* – converts to hollandite at high temperatures • breakage: no cleavage; uneven fracture • tenacity: brittle • habit: always in aggregates that are internally granular and externally massive, colloform, or stalactitic • luster: submetallic • diaphaneity: opaque • fusibility: 7 • other: yields some water condensate when heated in a test tube □ use: black paint pigment
black to brownish-black	black	4.72-4.79	5-6	**ilmenite, $FeTiO_3$** [a.k.a. menaccanite, mohsite, and titanic iron] *(hexagonal)* • breakage: no cleavage; conchoidal fracture • tenacity: brittle • habit: usually in aggregates that are internally finely granular and externally massive; and rarely in hexagonal tabular crystals • luster: metallic to submetallic • diaphaneity: pseudo-opaque to opaque • fusibility: 7 • other: moderately attracted to a neodymium magnet □ use: possibly white paint pigment when smelted to produce Ti oxide

black	black	5.17-5.20	5.5-6.5	**magnetite, Fe_3O_4** [a.k.a. magnetic iron] *(isometric)* • breakage: no cleavage but occasionally prominent parting in one direction; also sub-conchoidal fracture • tenacity: brittle • habit: commonly in octahedral and, less often, dodecahedral crystals; and also commonly in aggregates that are internally fine to coarse granular and externally massive • luster: metallic to submetallic • diaphaneity: opaque • fusibility: 5-5.5 • other: strongly attracted to neodymium and other weaker magnets, and occasionally can act as a magnet (= lodestone) □ use: smelted for iron; when lodestone, employed for compasses or to magnetize iron needles
dark-gray to pale-brown	usually reddish-brown; rarely black, bluish-violet, red, or yellow	4.18-5.50	6-6.5	**rutile, TiO_2** [a.k.a. red schorl] *(tetrahedral)* • breakage: cleavage in 4 directions (two prismatic), none prominent; also uneven to sub-conchoidal fracture • tenacity: brittle • habit: usually in acicular to slender prismatic crystals that have parallel longitudinal striations (commonly occurs as elbow twins and also as acicular inclusions within quartz crystals); and occasionally in aggregates that are internally compact granular and externally massive • luster: submetallic to nonmetallic adamantine • diaphaneity: transparent to pseudo-opaque • fusibility: 7 • other: sometimes triboluminescent □ use: possibly for jewelry and decorative arts, especially as 'rutilated' quartz

light-gray to light-brown or, when pure, white	usually reddish-brown to brownish-black; rarely yellow or, when pure, white	6.80-7.10	6-7	**cassiterite, SnO_2** [a.k.a. tinstone and ruby tin] *(tetragonal)* • breakage: cleavage in 3 directions (prismatic + basal pinacoidal), none prominent; also uneven to sub-conchoidal fracture • tenacity: brittle • habit: usually in aggregates that are internally granular or fibrous and externally massive or colloform; and occasionally in short prismatic and, less commonly, dipyramidal crystals • luster: submetallic to nonmetallic adamantine • diaphaneity: transparent to mainly translucent or pseudo-opaque • fusibility: 7 □ use: smelted for tin; possibly reduced to tin oxide (SnO) and used as a white paint pigment and glass colorant
C. Mohs Hardness > 5.5				
2. Streak shades of brown, yellow, or red				
i. Fusibility ≤ 3				
Specific Gravity > 2.70				
dark-brown	shiny black to brownish-black (tarnishing to dull black)	9.56	5.5	**plattnerite, PbO_2** *(tetragonal)* • breakage: no cleavage; sub-conchoidal to splintery fracture • tenacity: brittle • habit: usually in aggregates that are internally fibrous and externally colloform; and rarely in prismatic crystals • luster: metallic to submetallic • diaphaneity: opaque • fusibility: 2 □ use: possibly for brown paint pigment

<table>
<tr><th colspan="5">C. Mohs Hardness > 5.5</th></tr>
<tr><th colspan="5">2. Streak shades of brown, yellow, or red</th></tr>
<tr><th colspan="5">ii. Fusibility > 3</th></tr>
<tr><th colspan="5">Specific Gravity > 2.70</th></tr>
<tr><td rowspan="2">red to reddish-brown</td><td>gray to rarely black</td><td rowspan="2">4.90-5.26</td><td rowspan="2">5-6.5</td><td rowspan="2">hematite (or haematite), Fe_2O_3 (hexagonal)
<u>metallic variety:</u> specular hematite or specularite
• breakage: no cleavage; uneven to sub-conchoidal fracture
• tenacity: brittle to slightly elastic
• habit: usually in aggregates that are internally foliated and externally massive; and rarely in hexagonal tabular crystals
• luster: metallic
• diaphaneity: opaque
• fusibility: 7
• other: moderately to strongly attracted to a neodymium magnet
□ use: smelted for iron; red paint pigment and colorant for glass and vitreous glazes; small decorative objects; weights; medicine

<u>submetallic to nonmetallic variety:</u> common hematite or red ocher [a.k.a. bloodstone, red iron, and kidney stone]
• breakage: no cleavage; uneven fracture
• tenacity: brittle to friable or pulverulent
• habit: usually in aggregates that are internally fibrous or granular and externally massive, colloform, oolitic, pisolitic, or stalactitic
• diaphaneity: pseudo-opaque to opaque
• luster: submetallic to nonmetallic dull or silky
• fusibility: 7
• other: weakly to moderately attracted to a neodymium magnet; commonly has clay mineral impurities when ocherous
□ use: smelted for iron; red paint pigment and colorant for glass and vitreous glazes; small decorative objects; weights; medicine</td></tr>
<tr><td>red to brownish-red</td></tr>
</table>

dark-brown	black to brownish black (commonly with green Cr-oxide corrosion)	4.50-5.09	5.5-6	**chromite, $FeCr_2O_4$** [a.ka. chrome or chromic iron] *(isometric)* • breakage: no cleavage; even to mainly uneven fracture • tenacity: brittle • habit: usually in aggregates that are internally compact granular and externally massive; and rarely in octahedral crystals • luster: metallic to submetallic • diaphaneity: pseudo-opaque to opaque • fusibility: 7 • other: weakly attracted to a neodymium magnet □ use: green colorant in vitreous glazes
dark-gray to pale-brown	usually reddish-brown; rarely black, bluish-violet, red, or yellow	4.18-5.50	6-6.5	**rutile, TiO_2** [a.k.a. red schorl] *(tetrahedral)* • breakage: cleavage in 4 directions (two prismatic), none prominent; also uneven to sub-conchoidal fracture • tenacity: brittle • habit: usually in acicular to slender prismatic crystals that have parallel longitudinal striations (commonly occurs as elbow twins and also as acicular inclusions within quartz crystals); and occasionally in aggregates that are internally compact granular and externally massive • luster: submetallic to nonmetallic adamantine • diaphaneity: transparent to pseudo-opaque • fusibility: 7 • other: sometimes triboluminescent □ use: possibly for jewelry and decorative arts, especially as 'rutilated' quartz
light-gray to light-brown or, when pure, white	usually reddish-brown to brownish-black; rarely yellow or, when pure, white	6.80-7.10	6-7	**cassiterite, SnO_2** [a.k.a. tinstone and ruby tin] *(tetragonal)* • breakage: cleavage in 3 directions (prismatic + basal pinacoidal), none prominent; also uneven to sub-conchoidal fracture • tenacity: brittle • habit: usually in aggregates that are internally granular or fibrous and externally massive or colloform; and occasionally in short prismatic and, less commonly, dipyramidal crystals • luster: submetallic to nonmetallic adamantine • diaphaneity: transparent to mainly translucent or pseudo-opaque • fusibility: 7 □ use: smelted for tin; possibly reduced to tin oxide (SnO) and used as a white paint pigment and glass colorant

Table 3.5: Determination of archaeologically significant and common rock-forming minerals with a nonmetallic luster and a colored (non-white) streak.

II. LUSTER = NONMETALLIC & STREAK = COLORED (NON-WHITE)				
A. Mohs Hardness ≤ 2.5				
1. Streak shades of brown, yellow, or red				
i. Fusibility ≤ 3				
a. Specific Gravity ≤ 2.70				
Streak Color	Mineral Color	Specific Gravity	Mohs Hardness	Mineral Name and Other Properties
pale-yellow to yellowish-brown	yellow to yellowish-brown (occasionally with a greenish or reddish tinge)	2.05-2.09	1.5-2.5	**native sulfur**, S [a.k.a. brimstone] *(orthorhombic)* • breakage: no cleavage; uneven to conchoidal fracture • tenacity: brittle to slightly sectile • habit: usually in dipyramidal or tabular crystals; and occasionally in aggregates that are internally granular and externally massive, reniform, or stalactitic • luster: nonmetallic adamantine or resinous to greasy • diaphaneity: translucent to transparent • fusibility: 1 • other: when heated in a test tube melts, emits a strong sulfurous odor, and leaves a dark-red residue that cools to light-yellow; decrepitates and burns in a blowpipe or candle flame; colors a blowpipe flame a conspicuous blue; produces a sulfurous smell when freshly crushed or powdered; when held in a warm hand near the ear it may be heard to crack due to heat-induced expansion □ use: an ingredient in gun powder and incendiary weapons; yellow paint pigment; medicine; disinfectant and fumigant; cleaning cloth and bleaching wool; small decorative objects; filler in hollow gold jewelry
A. Mohs Hardness ≤ 2.5				
1. Streak shades of brown, yellow, or red				
i. Fusibility ≤ 3				
b. Specific Gravity > 2.70				
pale-yellow or white	colorless, white to yellowish-white, gray to yellowish-gray, or brown (darkening upon exposure to sunlight)	6.48-7.15	1-2	**calomel**, Hg_2Cl_2 [a.k.a. calomelite, horn mercury, and horn quicksilver] *(tetragonal)* • breakage: cleavage in 2 directions (prismatic), none prominent; also conchoidal fracture • tenacity: sectile • habit: usually in small prismatic, tabular, or dipyramidal crystals; and occasionally in aggregates that are internally granular and externally massive • luster: nonmetallic adamantine • diaphaneity: transparent • fusibility: 0 (volatizes) • other: commonly fluoresces pink or red in short- and long-wavelength UV lights □ use: possibly smelted for mercury

<table>
<tr><td>orangish-red</td><td>red to orangish-yellow</td><td>3.48-3.56</td><td>1.5-2</td><td>realgar, AsS [a.k.a. red arsenic] (monoclinic)
• breakage: cleavage in 2 directions (both pinacoidal), none prominent; also uneven to conchoidal fracture
• tenacity: sectile
• habit: usually in aggregates that are internally fine to coarse granular and externally massive; and occasionally in short prismatic crystals with parallel longitudinal striations
• luster: nonmetallic resinous to greasy
• diaphaneity: transparent (crystals) to translucent (aggregates)
• fusibility: 1
• other: when heated in a test tube emits a garlic-like odor (AsH_3), boils, leaves a shiny black residue of metallic arsenic, and also leaves an initially red then, with more heating, black sublimate of arsenic sulfide; colors a blowpipe flame blue but this can be difficult to see
□ use: bright red to orangish-red paint and cosmetic pigments; possibly smelted for arsenic or added directly to copper to manufacture arsenical bronze</td></tr>
<tr><td>pale-yellow</td><td>yellow to brownish-yellow</td><td>3.40-3.49</td><td>1.5-2</td><td>orpiment, As_2S_3 [a.k.a. yellow arsenic and rat's bane] (monoclinic)
• breakage: cleavage in 2 directions (both pinacoidal), one prominent
• tenacity: sectile, flexible but not elastic
• habit: usually in aggregates that are internally fibrous or foliated and externally massive or stalactitic; and rarely in tabular or short prismatic crystals
• luster: nonmetallic resinous, pearly, or dull
• diaphaneity: transparent to mainly translucent
• fusibility: 1
• other: when heated in a test tube emits a garlic-like odor (AsH_3), boils, leaves a shiny black residue of metallic arsenic, and also leaves an initially yellow then, with more heating, black sublimate of arsenic sulfide; colors a blowpipe flame blue but this can be difficult to see
□ use: bright yellow paint and cosmetic pigments; possibly smelted for arsenic or added directly to copper to manufacture arsenical bronze</td></tr>
</table>

bright red	brownish-red when impure to bright red when pure	8.09-8.10	2-2.5	**cinnabar, HgS** [a.k.a. vermillion] *(hexagonal)* • breakage: cleavage in 3 directions (rhombohedral), all prominent; also uneven to sub-conchoidal fracture • tenacity: slightly sectile • habit: commonly in hexagonal prismatic to tabular and rhombohedral crystals; and also commonly in aggregates that are internally fine granular and externally massive • luster: submetallic to mainly nonmetallic adamantine (crystals) or dull (aggregates) • diaphaneity: transparent to translucent • fusibility: 0 (volatizes) • other: when heated in a test tube volatizes, giving off toxic mercuric fumes and yielding a black sublimate of mercuric sulfide that becomes red when rubbed after cooling □ use: red cosmetic and paint pigments; jewelry and small decorative objects; gilding when Hg is mixed with gold and silver; smelted for mercury; medicine
A. Mohs Hardness ≤ 2.5				
1. Streak shades of brown, yellow, or red				
ii. Fusibility not determined				
Specific Gravity > 2.70				
light-yellow	light-yellow to orangish-brown	3.52	1-1.5	**pararealgar, As_4S_4** *(monoclinic)* • breakage: no cleavage; uneven fracture • tenacity: brittle to pulverulent • habit: almost always in aggregates that are internally fine granular and externally massive • luster: nonmetallic vitreous to resinous • diaphaneity: translucent to opaque • fusibility: not determined □ use: yellow paint pigment
A. Mohs Hardness ≤ 2.5				
2. Streak shades of blue or green				
i. Fusibility ≤ 3				
a. Specific Gravity ≤ 2.70				
pale-blue to bluish-white	colorless, white, or green (tarnishing to shades of blue)	2.66-2.71	1.5-2	**vivianite, $Fe_3(PO_4)_2 \cdot 8H_2O$** [a.k.a. blue iron earth and blue ocher] *(monoclinic)* • breakage: cleavage in 1 direction (pinacoidal), prominent • tenacity: flexible and slightly sectile; sometimes pulverulent • habit: commonly in bladed prismatic crystals that occur in radiated groups; and also commonly in aggregates that are internally fibrous and externally colloform • luster: nonmetallic vitreous to pearly • diaphaneity: transparent to pseudo-opaque • fusibility: 2-2.5 • other: yields brownish water when heated in a test tube; fuses to a black magnetic globule; colors a blowpipe flame bluish-green but this can be difficult to see; moderately attracted to a neodymium magnet □ use: blue paint pigment

A. Mohs Hardness ≤ 2.5				
2. Streak shades of blue or green				
i. Fusibility ≤ 3				
b. Specific Gravity > 2.70				
pale-blue	light-green, bluish-green, greenish-blue, or blue	3.00-3.25	1.5-2	**tyrolite, $Ca_2Cu_9(AsO_4)_4(CO_3)(OH)_8{\cdot}11H_2O$** [a.k.a. aphrochalcite , trichalcite, and copper froth] *(monoclinic)* • breakage: cleavage in 1 direction (pinacoidal), prominent • tenacity: brittle to slightly sectile • habit: commonly in acicular prismatic crystals that occur in radiated groups; and also commonly in aggregates that are internally fibrous and externally botryoidal • luster: nonmetallic vitreous to pearly or silky • diaphaneity: transparent to translucent • fusibility: 2-2.5 • other: effervesces strongly in cold dilute HCl acid □ use: possibly smelted for copper
A. Mohs Hardness ≤ 2.5				
2. Streak shades of blue or green				
ii. Fusibility > 3				
a. Specific Gravity ≤ 2.70				
pale-green	usually green, bluish-green or greenish-black; rarely colorless or yellowish-green	2.40-2.95	2	**glauconite, $(K,Na)(Fe,Al,Mg)_2(Si,Al)_4O_{10}(OH)_2$** [mica group] *(monoclinic)* • breakage: cleavage in 1 direction (basal pinacoidal), prominent • tenacity: brittle • habit: almost always in aggregates that are internally foliated and externally in small rounded grains • luster: nonmetallic dull • diaphaneity: pseudo-opaque • fusibility: ~7 • other: when heated in a test tube yields water condensate; weakly attracted to a neodymium magnet but becomes moderately to strongly attracted after heating in a candle or blowpipe flame □ use: green paint pigment and colorant in vitreous glazes

A. Mohs Hardness ≤ 2.5				
2. Streak shades of blue or green				
ii. Fusibility > 3				
b. Specific Gravity > 2.70				
light-blue	pale-green to light- or greenish-blue	3.93-3.96	1-2	**aurichalcite, $(Zn,Cu)_5(CO_3)_2(OH)_6$** [a.k.a. blue calamine and brass ore] *(monoclinic or orthorhombic depending on crystallization conditions)* • breakage: cleavage in 1 direction (pinacoidal), prominent; also uneven fracture • tenacity: brittle • habit: usually in acicular to slender bladed prismatic crystals, commonly occurring in plumose tufts; and rarely in aggregates that are internally granular and externally stalactitic • luster: nonmetallic silky to pearly • diaphaneity: transparent • fusibility: 7 • other: colors a blowpipe flame dark-green but this can be difficult to see; effervesces strongly in cold dilute HCl acid □ use: possibly smelted for both copper and zinc to make brass
pale-green	usually green, bluish-green or greenish-black; rarely colorless or yellowish-green	2.40-2.95	2	**glauconite, $(K,Na)(Fe,Al,Mg)_2(Si,Al)_4O_{10}(OH)_2$** [mica group] *(monoclinic)* • breakage: cleavage in 1 direction (basal pinacoidal), prominent • tenacity: brittle • habit: almost always in aggregates that are internally foliated and externally in small rounded grains • luster: nonmetallic dull • diaphaneity: pseudo-opaque • fusibility: ~7 • other: when heated in a test tube yields water condensate; weakly attracted to a neodymium magnet but becomes moderately to strongly attracted after heating in a candle or blowpipe flame □ use: green paint pigment and colorant in vitreous glazes
B. Mohs Hardness > 2.5 to ≤ 5.5				
1. Streak shades of brown, yellow, or red				
Fusibility > 3 and Specific Gravity > 2.70				
pale-yellow	brownish-yellow to dark-brown	3.25	2.5-3.5	**jarosite, $KFe_3(SO_4)_2(OH)_6$** [a.k.a. utahite] *(hexagonal)* • breakage: cleavage in 1 direction (basal pinacoidal), not prominent; also uneven to conchoidal fracture • tenacity: brittle • habit: usually in ocherous aggregates that are internally granular or fibrous and externally massive; and rarely in tabular or pseudo-cubical crystals • luster: nonmetallic vitreous to resinous or dull • diaphaneity: translucent to pseudo-opaque • fusibility: 4.5 • other: soluble in hot dilute HCl acid; strongly pyroelectric □ use: yellow paint pigment

pale-yellow	yellow or yellowish- to reddish-brown	2.90-3.30	2.5-3.5	**natrojarosite, $NaFe_3(SO_4)_2(OH)_6$** [a.k.a. utahite] *(hexagonal)* • breakage: cleavage in 1 direction (basal pinacoidal), not prominent; also uneven to conchoidal fracture • tenacity: brittle • habit: usually in aggregates that are internally granular or fibrous and externally massive; and rarely in hexagonal tabular or pseudo-cubical crystals • luster: nonmetallic vitreous • diaphaneity: transparent to translucent • fusibility: 4.5 □ use: yellow paint pigment
reddish-brown, light-yellow; rarely white when pure	commonly yellow, yellowish-brown, dark-brown or black; occasionally red, green, or, when pure, white or colorless	3.90-4.10	3.5-4	**sphalerite, ZnS** [a.k.a. blende, zinc blende, mock lead, and black or steel jack] *(isometric)* • breakage: cleavage in 6 directions (dodecahedral), all prominent; also conchoidal fracture • tenacity: brittle • habit: usually in poorly-formed, tetragonal or dodecahedral crystals; and occasionally in aggregates that are internally fibrous or fine to coarse granular and externally botryoidal • luster: submetallic to nonmetallic adamantine or resinous • diaphaneity: translucent to transparent • fusibility: 7 • other: gives off a sulfurous smell when crushed; when powdered, emits a rotten egg smell (from H_2S) in cold dilute HCl acid; pyroelectric and commonly triboluminescent □ use: smelted for zinc, an ingredient of brass
shiny brownish-red	dark-red to reddish-black	6.10-6.15	3.5-4	**cuprite, Cu_2O** [a.k.a. ruby copper] *(isometric)* • breakage: cleavage in 3 directions (cubic), none prominent; also uneven to conchoidal fracture • tenacity: brittle • habit: usually in cubical or octahedral crystals and less commonly in dodecahedral or filiform crystals; and occasionally in aggregates that are internally granular and externally massive • luster: submetallic to nonmetallic adamantine (crystals) or dull (aggregates) • diaphaneity: pseudo-opaque • fusibility: 3 • other: colors a blowpipe flame dark-green but this can be difficult to see □ use: smelted for copper

yellow to yellowish-brown	usually yellow to yellowish-brown; sometimes other shades of brown	2.70-4.30	4-5.5	**limonite, $\sim FeO(OH) \cdot H_2O$** [a.k.a. brown and yellow ochers, brown hematite or iron, and bog iron] *(mineraloid or possibly an intergrowth of goethite and lepidocrocite)* • breakage: no cleavage; uneven fracture • tenacity: friable to brittle • habit: always in soft (earthy or ocherous) to hard aggregates that are internally fibrous and externally massive, mammillary, or stalactitic • luster: submetallic to nonmetallic dull or silky • diaphaneity: translucent to opaque • fusibility: 5-5.5 • other: yields water condensate when heated in test tube, dehydrating to hematite; weakly to moderately attracted to a neodymium magnet; commonly has clay mineral impurities when ocherous □ use: smelted for iron; yellow to brownish-yellow paint pigments and colorants in glass and vitreous glazes
brownish- to orangish-yellow or yellowish-brown	dark-brown, reddish- to yellowish-brown, brownish-yellow, and brownish-black	3.30-4.37	5-5.5	**goethite, $\alpha FeO(OH)$** [a.k.a. allcharite, bog iron, and xanthosiderite] *(orthorhombic)* • breakage cleavage in 2 directions (both pinacoidal), one prominent; also uneven fracture • tenacity: brittle to pulverulent • habit: usually in aggregates that are internally fibrous to compact granular and externally massive, botryoidal, reniform, oolitic/pisolitic, or stalactitic; and rarely in acicular prismatic, tabular, or filiform crystals (but commonly pseudomorphs after cubical pyrite crystals) • luster: submetallic to nonmetallic dull or silky (aggregates), or adamantine (crystals) • diaphaneity: pseudo-opaque to opaque • fusibility: 5.5 • other: when heated in a test tube yields water condensate, dehydrating to hematite; weakly to moderately attracted to a neodymium magnet □ use: yellow to brownish-yellow paint pigments and colorants in glass and vitreous glazes; small decorative objects; smelted for iron

red to reddish-brown	gray to rarely black	4.90-5.26	5-6.5	**hematite** (or **haematite**), Fe_2O_3 *(hexagonal)* metallic variety: **specular hematite** or **specularite** • breakage: no cleavage; uneven to sub-conchoidal fracture • tenacity: brittle to slightly elastic • habit: usually in aggregates that are internally foliated and externally massive; and rarely in hexagonal tabular crystals • luster: metallic • diaphaneity: opaque • fusibility: 7 • other: moderately to strongly attracted to a neodymium magnet □ use: smelted for iron; red paint pigment and colorant for glass and vitreous glazes; small decorative objects; weights; medicine
	red to brownish-red			submetallic to nonmetallic variety: **common hematite** or **red ocher** [a.k.a. bloodstone, red iron, and kidney ore] • breakage: no cleavage; uneven fracture • tenacity: brittle to friable or pulverulent • habit: usually in aggregates that are internally fibrous or granular and externally massive, colloform, oolitic, pisolitic, or stalactitic • diaphaneity: pseudo-opaque to opaque • luster: submetallic to nonmetallic dull or silky • fusibility: 7 • other: weakly to moderately attracted to a neodymium magnet; commonly has clay mineral impurities when ocherous □ use: smelted for iron; red paint pigment and colorant for glass and vitreous glazes; small decorative objects; weights; medicine
B. Mohs Hardness > 2.5 to ≤ 5.5				
2. Streak shades of blue or green				
i. Fusibility ≤ 3				
Specific Gravity > 2.70				
bluish-green	blue or greenish-blue to greenish-black	4.38-4.42	2.5-3	**clinoclase, $Cu_3(AsO_4)(OH)_3$** [a.k.a. clinoclasite] *(monoclinic)* • breakage: cleavage in 1 direction (pinacoidal), prominent; also uneven fracture • tenacity: brittle • habit: usually in prismatic, tabular, or pseudo-rhombohedral crystals that are commonly in roseiform groups; and occasionally in aggregates that are internally fibrous and externally massive • luster: nonmetallic vitreous to pearly • diaphaneity: transparent to translucent • fusibility: 2-2.5 □ use: possibly smelted for copper

brownish-, grayish- or yellowish-green	usually grayish-green or blackish-green; sometimes gray or yellowish-brown	4.10-4.46	3	**olivenite, $Cu_2AsO_4(OH)$** [a.k.a. leucochalcite and wood copper] *(monoclinic or orthorhombic depending on crystallization conditions)* • breakage: cleavage in 4 directions (two prismatic), none prominent; also uneven to conchoidal fracture • tenacity: brittle • habit: commonly in acicular or short, slender prismatic or tabular crystals that are in radiated groups; and also commonly in aggregates that are internally granular or fibrous and externally massive or colloform • luster: nonmetallic vitreous to resinous, pearly, or silky • diaphaneity: translucent to opaque • fusibility: 2-2.5 • other: colors a blowpipe flame bluish-green but this can be difficult to see; when powdered, fully dissolves in hot dilute HCl acid and leaves the solution light-greenish-blue, but in cold dilute HCl only partially dissolves and leaves the solution faintly green □ use: possibly smelted for copper; possibly green paint pigment
pale-green	bright green, commonly banded with alternating lighter and darker shades of green	3.90-4.03	3.5-4	**malachite, $Cu_2CO_3(OH)_2$** *(monoclinic)* • breakage: cleavage in 2 directions (both pinacoidal), one prominent but difficult to see; also uneven fracture • tenacity: brittle • habit: usually in aggregates that are internally compact granular or fibrous and externally massive or colloform; and occasionally in acicular to prismatic crystals that occur in roseiform groups • luster: nonmetallic adamantine to vitreous (crystals) or silky to dull (aggregates) • diaphaneity: translucent (crystals) to pseudo-opaque or opaque (aggregates) • fusibility: 3 • other: when heated in a test tube yields water condensate and leaves a black residue; decrepitates in a blowpipe flame and colors it dark-green but this can be difficult to see; in cold dilute HCl acid effervesces either weakly (5%) or strongly (10%) when whole, but strongly in both solutions when powdered □ use: green paint pigment and colorant in vitreous glazes; smelted for copper; jewelry and decorative arts

pale-blue	light- to dark-purplish-blue	3.77-3.89	3.5-4	**azurite, $Cu_3(CO_3)_2(OH)_2$** [a.k.a. chessylite and blue malachite] *(monoclinic)* • breakage: cleavage in 3 directions (prismatic + pinacoidal), two (prismatic) prominent; also conchoidal fracture • tenacity: brittle • habit: commonly in tabular or short prismatic crystals that are often malformed and in radiated groups; and also commonly in aggregates that are internally fibrous and externally massive, colloform, or stalactitic • luster: nonmetallic vitreous (crystals) to dull (aggregates) • diaphaneity: transparent to translucent (crystals) or opaque (aggregates) • fusibility: 3 • other: when heated in a test tube yields water condensate and leaves a black residue; colors a blowpipe flame dark-green but this can be difficult to see; effervesces strongly in cold dilute HCl acid when whole □ use: smelted for copper; blue paint pigment and colorant in vitreous glazes; jewelry and decorative arts
B. Mohs Hardness > 2.5 to ≤ 5.5				
2. Streak shades of blue or green				
ii. Fusibility > 3 or not determined				
a. Specific Gravity ≤ 2.70				
pale-green	green to greenish-blue (or brown to black when impure)	2.00-2.40	2-4	**chrysocolla, $\sim Cu_4H_4Si_4O_{10}(OH)_8{\bullet}nH_2O$** [a.k.a. bisbeeite] *(a mineraloid)* • breakage: no cleavage; conchoidal fracture • tenacity: brittle • habit: always in aggregates that are internally compact fibrous and externally massive • luster: nonmetallic vitreous to waxy or dull • diaphaneity: pseudo-opaque • fusibility: 7 • other: when heated in a test tube yields water condensate; decrepitates in a blowpipe flame, turns black and then brown, and colors the flame dark-green but this can be difficult to see □ use: smelted for copper; jewelry and decorative arts; possibly green paint pigment

pale-blue	usually dark-blue, greenish-blue, or purplish-blue; occasionally white, gray, or green	2.38-2.45 lazurite 2.44-2.50 haüyne	5-5.5 lazurite 5.5-6 haüyne	**lazurite, $(Na,Ca)_8(AlSiO_4)_6(SO_4,S,Cl,OH)$** to **haüyne, $Na_6Ca_2(AlSiO_4)_6(SO_4)_2$** in a solid-state series [sodalite group ; a.k.a. ultramarine for lazurite and hauynite for haüyne] *(isometric)* • gem variety: primary mineral in **lapis lazuli,** which also commonly has patches and veins of white calcite and small grains of shiny yellowish pyrite • breakage: cleavage in 6 directions (dodecahedral), none prominent; also uneven fracture • tenacity: brittle • habit: almost always in aggregates that are internally compact granular and externally massive • luster: nonmetallic vitreous • diaphaneity: translucent to pseudo-opaque • fusibility: 3.5 • other: fuses with intumescence □ use: jewelry and decorative arts; blue paint pigment (ultramarine)
B. Mohs Hardness > 2.5 to ≤ 5.5				
2. Streak shades of blue or green				
ii. Fusibility > 3 or not determined				
b. Specific Gravity > 2.70				
green to greenish-white	dark-green to greenish-blue or greenish-black	3.71-3.77	3	**clinoatacamite, $Cu_2(OH)_3Cl$** *(monoclinic)* • breakage: cleavage in 1 direction (pinacoidal), prominent; also uneven fracture • tenacity: brittle • habit: almost always in small pseudo-rhombohedral crystals • luster: nonmetallic adamantine to vitreous • diaphaneity: transparent to translucent • fusibility: not determined □ use: green paint pigment; possibly smelted for copper
light-green	dark-green to greenish-black	3.72-3.75	3	**paratacamite, $Cu_3(Cu,Zn)(OH)_6Cl_2$** *(hexagonal)* • breakage: cleavage in 2 directions (prismatic), none prominent; also uneven to conchoidal fracture • tenacity: brittle • habit: commonly in rhombohedral crystals; and also commonly in aggregates that are internally granular and externally massive • luster: nonmetallic vitreous (crystals) or dull (aggregates) • diaphaneity: translucent to opaque • fusibility: not determined □ use: smelted for copper; green paint pigment

green	various shades of green	3.75-3.87	3-3.5	**atacamite, $Cu_2Cl(OH)_3$** [a.k.a. remolinite] *(orthorhombic)* • breakage: cleavage in 3 directions (prismatic + pinacoidal), one prominent • tenacity: brittle • habit: usually in aggregates that are internally granular, fibrous, or banded and externally massive; and occasionally in tabular or slender prismatic crystals with the latter having parallel longitudinal striations • luster: nonmetallic adamantine to vitreous • diaphaneity: translucent to pseudo-opaque • fusibility: 3-4 • other: when heated in a test tube yields water condensate and forms a gray sublimate; decrepitates in a blowpipe flame and colors it purplish-blue or green but this can be difficult to see; when powdered, dissolves in hot dilute HCl acid □ use: green paint pigment; possibly smelted for copper
pale-green	green to blackish-green	3.97	3.5-4	**brochantite, $Cu_4SO_4(OH)_6$** [a.k.a. blanchardite, kamarezite, and waringtonite] *(monoclinic)* • breakage: cleavage in 1 direction (pinacoidal), prominent; also uneven to conchoidal fracture • tenacity: brittle • habit: commonly in groups of acicular prismatic crystals; and also commonly in aggregates that are internally granular and externally massive • luster: nonmetallic vitreous • diaphaneity: transparent to translucent • fusibility: 3.5 • other: colors a blowpipe flame a conspicuous green to greenish-blue; when heated in a test tube yields abundant water condensate; when powdered, dissolves in hot dilute HCl acid and colors the solution light-blue □ use: smelted for copper
pale-green	green to blackish-green	3.88-3.90	3.5-4	**antlerite, $Cu_3SO_4(OH)_4$** [a.k.a. vernadskite] *(orthorhombic)* • breakage: cleavage in 2 directions (both pinacoidal), one prominent • tenacity: brittle • habit: usually in aggregates that are internally granular or fibrous and externally massive or reniform; and rarely in short, slender prismatic crystals with parallel longitudinal striations • luster: nonmetallic vitreous • diaphaneity: translucent • fusibility: possibly within the 3-5 range • other: colors a blowpipe flame green but this can be difficult to see □ use: green paint pigment; possibly smelted for copper

pale-green	green, bluish-green, greenish-blue, or dark-blue	3.40-3.42	3.5-4	**veszelyite, $(Cu,Zn)_2Zn(PO_4)(OH)_3{\bullet}2H_2O$** [a.k.a. arakawaite] *(monoclinic)* • breakage: cleavage in 2 directions (prismatic), none prominent • tenacity: brittle • habit: commonly in tabular, short prismatic, or pseudo-rhombohedral crystals; and also commonly in aggregates that are internally granular and externally massive • luster: nonmetallic vitreous • diaphaneity: translucent • fusibility: ≤4 □ use: green paint pigment
green	dark-green to bluish-green	3.28-3.35	5	**dioptase, $Cu_6(Si_6O_{18}){\bullet}H_2O$** [a.k.a. emerald copper] *(hexagonal)* • breakage: cleavage in 3 directions (rhombohedral), all prominent; also uneven to conchoidal fracture • tenacity: brittle • habit: almost always in hexagonal prismatic or rhombohedral crystals • luster: nonmetallic adamantine to vitreous • diaphaneity: transparent to translucent • fusibility: 7 □ use: possibly jewelry
light-blue	blue	3.08-3.09	5	**cuprorivaite, $CaCuSi_4O_{10}$** *(tetragonal)* • breakage: cleavage in 1 direction (basal pinacoidal), prominent • tenacity: brittle • habit: almost always in tabular crystals • luster: nonmetallic vitreous • diaphaneity: transparent to translucent • fusibility: not determined □ use: possibly blue paint pigment
pale-blue to bluish-gray	light- to dark- blue or greenish- to purplish-blue	3.02-3.43 increasing with Fe content	5-6	**glaucophane, $Na_2Mg_3Al_2Si_8O_{22}(OH)_2$** to **riebeckite, $Na_2Fe_3^{2+}Fe_2^{3+}Si_8O_{22}(OH)_2$** in a solid-solution series [amphibole group; a.k.a. crocidolite, crossite, and blue asbestos] *(monoclinic)* • breakage: cleavage in 2 directions (prismatic), both prominent • tenacity: brittle • habit: usually in aggregates that are internally fibrous and externally massive (especially riebeckite); and rarely in acicular crystals (especially glaucophane) • luster: nonmetallic vitreous (crystals) to silky (aggregates) • diaphaneity: translucent • fusibility: 3-3.5 □ use: blue paint pigment

<table>
<tr><td colspan="5">C. Mohs Hardness > 5.5</td></tr>
<tr><td colspan="5">1. Streak shades of brown, yellow, or red</td></tr>
<tr><td colspan="5">Fusibility >3 and Specific Gravity > 2.70</td></tr>
<tr><td rowspan="2">red to reddish-brown</td><td>black to mainly gray</td><td rowspan="2">4.90-5.26</td><td rowspan="2">5-6.5</td><td>hematite (or haematite), Fe_2O_3 (hexagonal)
<u>metallic variety:</u> specular hematite or specularite
• breakage: no cleavage; uneven to sub-conchoidal fracture
• tenacity: brittle to slightly elastic
• habit: usually in aggregates that are internally foliated and externally massive; and rarely in hexagonal tabular crystals
• luster: metallic
• diaphaneity: opaque
• fusibility: 7
• other: moderately to strongly attracted to a neodymium magnet
□ use: smelted for iron; red paint pigment and colorant for glass and vitreous glazes; small decorative objects; weights; medicine</td></tr>
<tr><td>red to brownish-red</td><td><u>submetallic to nonmetallic variety:</u> common hematite or red ocher [a.k.a. bloodstone, red iron, and kidney ore]
• breakage: no cleavage; uneven fracture
• tenacity: brittle to friable or pulverulent
• habit: usually in aggregates that are internally fibrous or granular and externally massive, colloform, oolitic, pisolitic, or stalactitic
• diaphaneity: pseudo-opaque to opaque
• luster: submetallic to nonmetallic dull or silky
• fusibility: 7
• other: weakly to moderately attracted to a neodymium magnet; commonly has clay mineral impurities when ocherous
□ use: smelted for iron; red paint pigment and colorant for glass and vitreous glazes; small decorative objects; weights; medicine</td></tr>
<tr><td>dark-gray to pale-brown</td><td>usually reddish-brown; rarely black, bluish-violet, red, or yellow</td><td>4.18-5.50</td><td>6-6.5</td><td>rutile, TiO_2 [a.k.a. ruby schorl] (tetrahedral)
• breakage: cleavage in 4 directions (two prismatic), none prominent; also uneven to sub-conchoidal fracture
• tenacity: brittle
• habit: usually in acicular to slender prismatic crystals that have parallel longitudinal striations (commonly occurs as elbow twins and also as acicular inclusions within quartz crystals); and occasionally in aggregates that are internally compact granular and externally massive
• luster: submetallic to nonmetallic adamantine
• diaphaneity: transparent to pseudo-opaque
• fusibility: 7
• other: sometimes triboluminescent
□ use: possibly for jewelry and decorative arts, especially as ‘rutilated’ quartz</td></tr>
</table>

light-gray to light-brown or, when pure, white	usually reddish-brown to brownish-black; rarely yellow or, when pure, white	6.80-7.10	6-7	**cassiterite, SnO_2** [a.k.a. tinstone and ruby tin] *(tetragonal)* • breakage: cleavage in 3 directions (prismatic + basal pinacoidal), none prominent; also uneven to sub-conchoidal fracture • tenacity: brittle • habit: usually in aggregates that are internally granular or fibrous and externally massive or colloform; and occasionally in short prismatic and, less commonly, dipyramidal crystals • luster: submetallic to nonmetallic adamantine • diaphaneity: transparent to mainly translucent or pseudo-opaque • fusibility: 7 □ use: smelted for tin; possibly reduced to tin oxide (SnO) and used as a white paint pigment and glass colorant
C. Mohs Hardness > 5.5				
2. Streak shades of gray or black				
Fusibility > 3 and Specific Gravity > 2.70				
dark-gray to pale-brown	usually reddish-brown; rarely black, bluish-violet, red, or yellow	4.18-5.50	6-6.5	**rutile, TiO_2** [a.k.a. ruby schorl] *(tetrahedral)* • breakage: cleavage in 4 directions (two prismatic), none prominent; also uneven to sub-conchoidal fracture • tenacity: brittle • habit: usually in acicular to slender prismatic crystals that have parallel longitudinal striations (commonly occurs as elbow twins and also as acicular inclusions within quartz crystals); and occasionally in aggregates that are internally compact granular and externally massive • luster: submetallic to nonmetallic adamantine • diaphaneity: transparent to pseudo-opaque • fusibility: 7 • other: sometimes triboluminescent □ use: possibly for jewelry and decorative arts, especially as 'rutilated' quartz
light-gray to light-brown or, when pure, white	usually reddish-brown to brownish-black; rarely yellow or, when pure, white	6.80-7.10	6-7	**cassiterite, SnO_2** [a.k.a. tinstone and ruby tin] *(tetragonal)* • breakage: cleavage in 3 directions (prismatic + basal pinacoidal), none prominent; also uneven to sub-conchoidal fracture • tenacity: brittle • habit: usually in aggregates that are internally granular or fibrous and externally massive or colloform; and occasionally in short prismatic and, less commonly, dipyramidal crystals • luster: submetallic to nonmetallic adamantine • diaphaneity: transparent to mainly translucent or pseudo-opaque • fusibility: 7 □ use: smelted for tin; possibly reduced to tin oxide (SnO) and used as a white paint pigment and glass colorant

Table 3.6: Determination of archaeologically significant and common rock-forming minerals with a nonmetallic luster, and either a white or white grading to pale-colored streak.

III. LUSTER = NONMETALLIC & STREAK = WHITE (+ PALE-COLORS)				
A. Mohs Hardness ≤ 2.5				
1. Fusibility ≤ 3				
a. Specific Gravity ≤ 2.70				
Streak Color	**Mineral Color**	**Specific Gravity**	**Mohs Hardness**	**Mineral Name and Other Properties**
white	colorless, white, or gray (commonly with yellowish tinges)	1.42-1.48	1-1.5	**natron, $Na_2CO_3 \cdot 10H_2O$** *(monoclinic)* • breakage: cleavage in 1 direction (basal pinacoidal), prominent; also conchoidal fracture • tenacity: brittle • habit: usually in aggregates that are internally granular and externally massive; and rarely in tabular crystals • luster: nonmetallic vitreous to dull • diaphaneity: transparent to translucent • fusibility: 1 • other: when heated in a test tube yields water condensate; colors a blowpipe flame a conspicuous vivid yellow; effervesces strongly in cold dilute HCl acid; bitter taste; soluble in water □ use: ingredient in glass, faience, and vitreous glazes; desiccant and defatting material in mummification; medicine; bleach for linen; food preparation
white	colorless, white, yellow, or blue	2.08-2.11	2	**niter** or **nitre, KNO_3** [a.k.a. saltpeter] *(orthorhombic)* • breakage: cleavage in 3 directions (all pinacoidal), one prominent • tenacity: brittle • habit: almost always in acicular crystals occurring in tuft-like radiated groups • luster: nonmetallic vitreous • diaphaneity: translucent to pseudo-opaque • fusibility: 1 • other: colors a blowpipe flame a conspicuous violet; soluble in water; salty taste □ use: gunpowder; metallurgical flux

white	usually colorless or white; occasionally pale-blue or violet	1.75-1.78	2-2.5	**alum group, $XAl_m(SO_4)_n\cdot{}_qH_2O$** where X = K, Co, Fe, Mg, Na or NH4; and m, n and q = variable integer amounts *(monoclinic)* • <u>common variety</u>: **potash-** or **K-alum, $KAl(SO_4)\cdot 12H_2O$** [a.k.a. kalinite] • breakage: cleavage 1 direction (pinacoidal), not prominent; also conchoidal fracture • tenacity: brittle • habit: usually in aggregates that are internally granular and externally massive; and occasionally in acicular crystals, and rarely in cubical or octahedral crystals • luster: nonmetallic vitreous • diaphaneity: transparent to translucent • fusibility: 1 • other: colors a blowpipe flame a conspicuous violet (K-alum); quickly dissolves in warm water; has an astringent taste □ use: desiccant in mummification; preservative for tawing hides; mordant/fixative for dyeing linen and leather; fumigant for wood; insect repellent; flocculent in wine to remove impurities; ingredient in glass and faience; when Co-bearing, a blue colorant in glass, vitreous glazes, and paint; deodorant; astringent to staunch blood flow; emetic to induce vomiting
white	colorless to white; sometimes tinged with gray, blue, or green	1.71	2-2.5	**borax, $Na_2B_4O_5(OH)_4\cdot 8H_2O$** [a.k.a. tincal] *(monoclinic)* • breakage: cleavage in 3 directions (prismatic + pinacoidal), one prominent; also conchoidal fracture • tenacity: brittle • habit: usually in tabular to short prismatic crystals; and occasionally in aggregates that are internally granular and externally massive • luster: nonmetallic vitreous to dull • diaphaneity: translucent to pseudo-opaque • fusibility: 1-1.5 • other: fuses in a blowpipe flame with intumescence and colors the flame yellow to yellowish-green; when heated in a test tube swells and blackens, yields abundant water condensate, and emits a burnt odor; dissolves in water; has a sweetish bitter taste; commonly phosphoresces blue or green in short- and long-wavelength UV lights □ use: ingredient in vitreous glazes and in solder for copper; eye antiseptic and salve

white	colorless or white	1.68	2-2.5	**epsomite, $MgSO_4 \cdot 7H_2O$** [a.k.a. bitter, hair or Epsom salts, and seelandite] *(orthorhombic)* • breakage: cleavage in 3 directions (prismatic + pinacoidal), one prominent • tenacity: brittle • habit: usually in aggregates that are internally granular or fibrous and externally colloform; and rarely in acicular prismatic or filiform crystals • luster: nonmetallic vitreous to dull or silky • diaphaneity: translucent to pseudo-opaque • fusibility: 1 • other: decrepitates in a blowpipe flame; dissolves in its own water upon heating in a test tube; has a bitter metallic taste □ use: medicine
white	usually colorless or white; when impure can be gray, yellow, red, blue, or purple	2.16-2.20	2-2.5	**halite, NaCl** [a.k.a. rock or common salt] *(isometric)* • breakage: cleavage in 3 directions (cubic), all prominent; also conchoidal fracture • tenacity: brittle • habit: usually in cubical crystals that commonly have recessed faces ('hopper-shaped'), also rarely in octahedral crystals; and occasionally in aggregates that are internally granular or rarely fibrous and externally massive or rarely stalactitic • luster: nonmetallic vitreous to greasy • diaphaneity: transparent to translucent • fusibility: 1.5 • other: decrepitates in a candle flame; colors a blowpipe flame a conspicuous orangish-yellow; salty taste; soluble in water, especially when warm; isotropic; sometimes triboluminescent □ use: food seasoning and preservative; preservative for hides; medicine; gold refining; ingredient is vitreous glazes
white	various shades of blue, especially light-blue	2.12-2.30	2.5	**chalcanthite, $CuSO_4 \cdot 5H_2O$** [a.k.a. blue vitriol] *(triclinic)* • breakage: cleavage in 1 direction (pinacoidal), not prominent; also conchoidal fracture • tenacity: brittle • habit: always in aggregates that are internally granular and externally massive, reniform, or stalactitic • luster: nonmetallic vitreous • diaphaneity: semi-transparent to translucent • fusibility: 3 • other: decrepitates in a blowpipe flame and colors it green; in a heated test tube yields water condensate and sulfurous fumes, and leaves a white powdery residue; very soluble in water, yielding a blue solution; has a nauseous metallic taste □ use: smelted for copper

A. Mohs Hardness ≤ 2.5				
1. Fusibility ≤ 3				
b. Specific Gravity > 2.70				
white to pale-yellow	colorless, white to yellowish-white, gray to yellowish-gray, or brown (darkening upon exposure to sunlight)	6.48-7.15	1-2	**calomel, Hg_2Cl_2** [a.k.a. calomelite, horn mercury, and horn quicksilver] *(tetragonal)* • breakage: cleavage in 2 directions (prismatic), none prominent; also conchoidal fracture • tenacity: sectile • habit: usually in small prismatic, tabular, or dipyramidal crystals; and occasionally in aggregates that are internally granular and externally massive • luster: nonmetallic adamantine • diaphaneity: transparent • fusibility: 0 (volatizes) • other: commonly fluoresces pink or red in short- and long-wavelength UV lights □ use: possibly smelted for mercury
shiny white	colorless when pure; shades of gray or bluish- to greenish-gray when impure (tarnishing to pale- to dark-purplish-brown in sunlight)	5.50	2-3	**chlorargyrite, AgCl** [a.k.a. cerargyrite and horn silver] *(isometric)* • breakage: no cleavage; uneven to subconchoidal fracture • tenacity: sectile • habit: usually in aggregates that are internally granular and externally massive; and rarely in cubical crystals • luster: nonmetallic waxy • diaphaneity: transparent to pseudo-opaque • fusibility: 1 • other: emits an acrid odor of chlorine when heated in a flame; isotropic when transparent or semi-transparent □ use: possibly smelted for silver

A. Mohs Hardness ≤ 2.5				
2. Fusibility > 3 or not determined				
a. Specific Gravity ≤ 2.70				
white	usually pale-green but commonly also white, greenish-gray, or dark-green (but commonly stained by iron oxides)	2.58-2.83	1	**talc, $Mg_3Si_4O_{10}(OH)_2$** [a.k.a. steatite, potstone, and soapstone when occurring as an impure, massive aggregate] *(monoclinic or triclinic depending on crystallization conditions)* • breakage: cleavage in 1 direction (basal pinacoidal), prominent with easy separation • tenacity: flexible but not elastic • habit: usually in aggregates that are internally compact granular or foliated and externally massive; and rarely in pseudo-hexagonal or pseudo-rhombohedral tabular crystals • luster: nonmetallic pearly to greasy or dull • diaphaneity: translucent • fusibility: 5-5.5 • other: decrepitates and whitens in a blowpipe flame; greasy feel □ use: filler and white pigment for paint; as steatite/soapstone - vessels, jewelry and small decorative objects, commonly with a colored, vitreous glaze
white	white, light-gray, or light-yellowish-brown (but commonly stained by iron oxides)	2.60-2.90	1-2	**illite, $K_{1-1.5}Al_4(Si_{6.5-7}Al_{1-1.5}O_{20})(OH)_4$** [clay group; a.k.a. hydromuscovite] *(monoclinic)* • breakage: cleavage in 1 direction (basal pinacoidal), prominent (but microscopic) with easy separation • tenacity: friable • habit: almost always in aggregates that are internally compact foliated and externally massive • luster: nonmetallic silky to dull • diaphaneity: translucent to pseudo-opaque • fusibility: 7 □ use: ceramics

white	usually gray, greenish-gray, or brownish-yellow; occasionally white, yellow, yellowish-green, pink, or brown	2.00-2.70	1.5-2.5	**montmorillonite, $(Na,Ca)_{0.3}(Al,Mg)_2Si_4O_{10}$•$12H_2O$** [clay group; a.k.a. smectite or, when occurring as an impure massive aggregate, bentonite and fuller's earth] *(monoclinic)* • breakage: cleavage in 1 direction (basal pinacoidal), prominent (but microscopic) with easy separation • tenacity: friable • habit: almost always in aggregates that are internally compact foliated and externally massive • luster: nonmetallic greasy to dull • diaphaneity: translucent to pseudo-opaque • fusibility: 7 • other: yields water condensate when heated in a test tube; greasy feel; greatly swells when wetted □ use: ceramics; cleaning cloth and degreasing wool (as fuller's earth); soap/washing powder; medicine
white	colorless, white, or light-gray (but commonly stained by iron oxides)	2.30-2.37	2	**gypsum, $CaSO_4$•$2H_2O$** *(monoclinic)* • <u>varieties</u>: (1) **selenite** – transparent, colorless crystals that break into thin sheets; (2) **satin spar** – a fibrous aggregate with a silky luster and chatoyancy; (3) **alabaster** – a moderately hard massive granular aggregate; and (4) **gypsite** – a soft and pulverulent, impure massive aggregate • breakage: cleavage in 4 directions (prismatic + two pinacoidal), one prominent; also conchoidal and splintery fracture • tenacity: sectile and flexible but not elastic • habit: commonly in tabular to long prismatic crystals that are bent or curved; and also commonly in aggregates that are internally compact to coarse granular, fibrous, or foliated and externally massive • luster: nonmetallic vitreous to pearly or silky • diaphaneity: transparent to translucent • fusibility: 3-3.5 • other: decrepitates in a candle flame; quickly yields water condensate when heated in a test tube, dehydrating to anhydrite; after dehydration, makes a hard plaster when water is added; colors a blowpipe flame yellowish-red but this can be difficult to see; quickly dissolves in hot dilute HCl acid; commonly fluoresces pale-yellow or bluish-white in short-wavelength, and pale-yellow or yellowish-white in long-wavelength UV lights □ use: plaster and mortar after calcining; an ingredient in vitreous glazes; white paint pigment; window panes for the alabaster and selenite varieties

white	white or pale-gray (but commonly stained by iron oxides)	2.61-2.68	2-2.5	**kaolinite, $Al_2Si_2O_5(OH)_4$** [clay group; a.k.a. kaolin] *(triclinic)* • breakage: cleavage in 1 direction (basal pinacoidal), prominent (but microscopic) with easy separation • tenacity: friable and pulverulent • habit: usually in aggregates that are internally compact foliated and externally massive; and rarely in minute pseudo-hexagonal tabular crystals • luster: nonmetallic dull (aggregates) or pearly (crystals) • diaphaneity: translucent to pseudo-opaque • fusibility: 7 • other: yields water when heated in a test tube; greasy feel; has an argillaceous or musty odor when breathed upon; will adhere to a dry tongue; commonly fluoresces white or yellowish-white in short- and long-wavelength UV lights □ use: ceramics; white paint pigment; medicine (internally for gastro-intestinal problems and externally for poultices)
white	usually white, pale-gray, or pale-yellow (sometimes with a faint pinkish tinge)	2.00-2.25	2-2.5	**sepiolite, $Mg_4(Si_6O_{15})(OH)_2 \cdot 6H_2O$** [clay group; a.k.a. meerschaum and sea-foam] *(mineraloid)* • breakage: no cleavage; splintery fracture • tenacity: brittle • habit: always in aggregates that are internally granular to mainly fibrous and externally massive; commonly micro-porous • luster: nonmetallic dull or silky • diaphaneity: translucent to pseudo-opaque • fusibility: 5 • other: yields water when heated in a test tube; decrepitates in a blowpipe flame and emits a burnt odor; floats on water when porous □ use: possibly a color-bearer for organic dyes in paint pigment; smoking pipes
white	usually white or pale-gray; occasionally pale-yellow, grayish-green, or brownish-white	2.10-2.60	2-2.5	**palygorskite, $(Mg,Al)_2Si_4O_{10}(OH) \cdot 4H_2O$** [clay group; a.k.a. attapulgite] *(monoclinic)* • breakage: cleavage in 1 direction (basal pinacoidal), not prominent: also uneven fracture • tenacity: tough • habit: usually in matted aggregates that are internally fibrous and externally massive; and rarely in small bladed crystals • luster: nonmetallic waxy or dull • diaphaneity: translucent • fusibility: not determined □ use: possibly a color-bearer for organic dyes in paint pigment

white to pale-green	usually various shades of green; rarely brown, yellow, pink, or red	2.6-3.3	2-3	**chlorite, $(Mg,Fe)_3(Si,Al)_4O_{10}(OH)_2$•$(Mg,Fe)_3(OH)_6$** [clay and mica groups; a.k.a. clinochlore, chamosite, and green mica] *(monoclinic or triclinic depending on crystallization conditions)* • breakage: cleavage in 1 direction (basal pinacoidal), prominent with easy separation • tenacity: flexible but not elastic • habit: usually in aggregates that are internally foliated and externally massive; and rarely in pseudo-hexagonal tabular crystals • luster: nonmetallic vitreous to pearly or dull • diaphaneity: transparent to translucent • fusibility: 5-5.5 • other: yields water condensate when heated in a test tube; whitens in a blowpipe flame and fuses along the edges; greasy feel □ use: none known, but it is a common rock-forming mineral
A. Mohs Hardness ≤ 2.5				
2. Fusibility > 3 or not determined				
b. Specific Gravity > 2.70				
white	usually pale-green but commonly also white, greenish-gray, or dark-green (but commonly stained by iron oxides)	2.58-2.83	1	**talc, $Mg_3Si_4O_{10}(OH)_2$** [a.k.a. steatite, potstone, or soapstone when occurring as an impure, massive aggregate] *(monoclinic or triclinic depending on crystallization conditions)* • breakage: cleavage in 1 direction (basal pinacoidal), prominent with easy separation • tenacity: flexible but not elastic • habit: usually in aggregates that are internally compact granular or foliated and externally massive; and rarely in pseudo-hexagonal or pseudo-rhombohedral tabular crystals • luster: nonmetallic pearly to greasy or dull • diaphaneity: translucent • fusibility: 5-5.5 • other: decrepitates and whitens in a blowpipe flame; greasy feel □ use: filler and white pigment for paint; as steatite/soapstone - vessels, jewelry and small decorative objects, commonly with a colored, vitreous glaze

white	white	2.70-2.87	1-2	**huntite, $CaMg_3(CO_3)_4$** *(hexagonal)* • breakage: no cleavage; uneven or sub-conchoidal fracture • tenacity: brittle • habit: almost always in aggregates that are internally fibrous and externally massive • luster: nonmetallic dull • diaphaneity: translucent • fusibility: not determined • other: effervesces strongly in cold dilute HCl acid; commonly fluoresces yellowish-white in short- and long-wavelength UV lights □ use: white paint pigment
white	white, light-gray, or light-yellowish-brown (but commonly stained by iron oxides)	2.60-2.90	1-2	**illite, $K_{1-1.5}Al_4(Si_{6.5-7}Al_{1-1.5}O_{20})(OH)_4$** [clay group; a.k.a. hydromuscovite] *(monoclinic)* • breakage: cleavage in 1 direction (basal pinacoidal), prominent (but microscopic) with easy separation • tenacity: friable • habit: almost always in aggregates that are internally compact foliated and externally massive • luster: nonmetallic silky to dull • diaphaneity: translucent to pseudo-opaque • fusibility: 7 □ use: ceramics
white to pale-gray	usually dark-brown or black (weathers to yellow); rarely dark-green for biotite — usually pale-yellow or yellowish-brown; rarely greenish- to reddish-brown for phlogopite	2.70-3.30	2-3	**biotite, $K(Mg,Fe)_3(AlSi_3O_{10})(OH)_2$** to **phlogopite, $KMg_3(AlSi_3O_{10})(OH)_2$** in a solid-solution series [mica group; a.k.a. black mica for biotite, and amber, brown or bronze mica for phologopite] *(monoclinic)* • breakage: cleavage in 1 direction (basal pinacoidal), prominent with easy separation • tenacity: flexible and elastic • habit: commonly in pseudo-hexagonal tabular or short prismatic crystals; and also commonly in aggregates that are internally foliated and externally massive • luster: submetallic to mainly nonmetallic vitreous or pearly • diaphaneity: transparent to translucent • fusibility: 5 • other: yields water condensate when heated in a test tube; immediately whitens in a blowpipe flame with fusion along the edges; biotite is weakly attracted to a neodymium magnet □ use: none known, but these are common rock-forming minerals

white to pale-green	usually various shades of green; rarely brown, yellow, pink, or red	2.6-3.3	2-3	**chlorite, $(Mg,Fe)_3(Si,Al)_4O_{10}(OH)_2\bullet(Mg,Fe)_3(OH)_6$** [clay and mica groups; a.k.a. clinochlore, chamosite, and green mica] *(monoclinic or triclinic depending on crystallization conditions)* • breakage: cleavage in 1 direction (basal pinacoidal), prominent with easy separation • tenacity: flexible but not elastic • habit: usually in aggregates that are internally foliated and externally massive; and rarely in pseudo-hexagonal tabular crystals • luster: nonmetallic vitreous to pearly or dull • diaphaneity: transparent to translucent • fusibility: 5-5.5 • other: yields water condensate when heated in a test tube; whitens in a blowpipe flame and fuses along the edges; greasy feel □ use: none known, but it is a common rock-forming mineral
white	usually colorless or pale-yellow (weathers to golden yellow); rarely pale-brown, pale-green, or pale-red	2.70-2.90	2-3.5	**muscovite, $KAl_2(AlSi_3O_{10})(OH)_2$** [mica group; a.k.a. common, white or potash mica, mirror stone, and Muscovy glass] *(monoclinic)* • breakage: cleavage in 1 direction, prominent with easy separation • tenacity: flexible and elastic • habit: usually in coarse aggregates that are internally foliated and externally massive; and occasionally in pseudo-hexagonal tabular crystals • luster: nonmetallic vitreous to pearly or silky • diaphaneity: transparent to translucent • fusibility: 4-5 • other: whitens in a blowpipe flame and fuses along the edges; yields water condensate when heated in a test tube; thin flakes held up to a bright light commonly exhibit asterism □ use: jewelry and decorative arts; small windows; mirrors; glitter in paint

<table>
<tr><th colspan="5">B. Mohs Hardness > 2.5 to ≤ 5.5</th></tr>
<tr><th colspan="5">1. Fusibility ≤ 3</th></tr>
<tr><th colspan="5">Specific Gravity ≤ 2.70</th></tr>
<tr><td>shiny white</td><td>colorless when pure and shades of gray or bluish- to greenish-gray when impure (tarnishing to light- to dark-purplish-brown in sunlight)</td><td>5.50</td><td>2-3</td><td>chlorargyrite, AgCl [a.k.a. cerargyrite or horn silver] (isometric)
• breakage: no cleavage; uneven to subconchoidal fracture
• tenacity: sectile
• habit: usually in aggregates that are internally granular and externally massive; and rarely in cubical crystals
• luster: nonmetallic waxy
• diaphaneity: transparent to pseudo-opaque
• fusibility: 1
• other: emits an acrid odor of chlorine when heated in a flame; isotropic when transparent or semi-transparent
□ use: possibly smelted for silver</td></tr>
<tr><td>white</td><td>colorless to white or pale-gray; commonly tinged with green, yellow, or blue</td><td>6.20-6.38</td><td>2.5-3</td><td>anglesite, $PbSO_4$ [a.k.a. lead spar or vitriol] (orthorhombic)
• breakage: cleavage in 3 directions (prismatic + pinacoidal), none prominent; also conchoidal fracture
• tenacity: brittle
• habit: usually in aggregates that are internally compact granular and banded and externally massive or colloform; and rarely in dipyramidal, prismatic, or tabular crystals
• luster: adamantine to vitreous or resinous (crystals) or dull (aggregates)
• diaphaneity: transparent to mainly translucent to opaque
• fusibility: 2.5
• other: decrepitates in a blowpipe flame; commonly fluoresces white, yellowish-white or pale-yellow in short-wavelength, and yellow, orangish-yellow or yellowish-orange in long-wavelength UV lights
□ use: smelted for lead; possibly white paint pigment</td></tr>
</table>

white	colorless to white or gray; darker with impurities and commonly tinged with blue or green	6.56-6.58	3-3.5	**cerussite, $PbCO_3$** [aragonite group; a.k.a. ceruse, lead spar, and white lead] *(orthorhombic)* • breakage: cleavage in 4 directions (two prismatic), none prominent; also conchoidal fracture • tenacity: brittle • habit: commonly in tabular, acicular prismatic, or dipyramidal crystals that are often pseudo-hexagonal and in stellate and reticular groups; and also commonly in aggregates that are internally fibrous or compact granular and externally massive or stalactitic • luster: nonmetallic adamantine to, less often, vitreous or resinous • diaphaneity: transparent to pseudo-opaque • fusibility: 1.5 • other: quickly decrepitates in a candle or blowpipe flame; when heated in a test tube turns yellow, then darkens with continued heating but returns to yellow when cool; effervesces weakly in cold dilute HCl acid when whole but strongly if powdered; stained purple by Alizarin Red S dye; commonly fluoresces yellow or yellowish-white in long-wavelength UV light □ use: smelted for lead and also silver, a sometimes abundant impurity
white	usually white or light- to brownish-yellow; occasionally colorless, green, pink, red, or purple	4.32-4.48	3.5	**adamite, $Zn_2(AsO_4)(OH)$** [a.k.a. adamine] *(orthorhombic)* • breakage: cleavage in 3 directions (prismatic + pinacoidal), none prominent; also uneven to conchoidal fracture • tenacity: brittle • habit: commonly in prismatic crystals that are in radiated or plumose groups; and also commonly in aggregates that are internally fibrous and externally colloform • luster: nonmetallic vitreous, waxy, or greasy • diaphaneity: transparent to translucent • fusibility: 3 • other: decrepitates and, when colored, turns white in a blowpipe flame □ use: possibly smelted for zinc, an ingredient of brass

white	usually pale-yellow, orangish-yellow, yellowish-brown, orangish-red, or red; occasionally colorless, white, or green	7.24	3.5-4	**mimetite, $Pb_5(AsO_4)_3Cl$** [a.k.a. mimetene and mimetesite] *(hexagonal)* • breakage: no cleavage; uneven to sub-conchoidal fracture • tenacity: brittle • habit: usually in acicular to mainly short, rounded (barrel-shaped), hexagonal prismatic crystals; and occasionally in aggregates that are internally granular and externally colloform or stalactitic • luster: nonmetallic vitreous to resinous • diaphaneity: transparent to translucent • fusibility: 1.5 • other: produces a white sublimate when heated in a test tube □ use: possibly smelted for lead
white	a wide range of colors including colorless, white to pale-gray, pale-yellow, pale-green, bluish-green to greenish-blue, bluish-violet, pale-blue, purple, bluish-black, or brown with common color zoning	3.18 when pure, otherwise 3.00-3.25	4	**fluorite, CaF_2** [a.k.a. fluor, fluorspar, and Derbyshire spar] *(isometric)* • gem varieties: **blue john** – purple to bluish-violet, commonly with white veins; and **color-banded** varieties • breakage: cleavage in 4 directions (octahedral), all prominent; also conchoidal to uneven fracture • tenacity: brittle • habit: usually in cubical crystals and, less commonly, octahedral or dodecahedral crystals with cubical penetration twins common; and occasionally in aggregates that are internally compact to coarse granular and externally massive or stalactitic • luster: nonmetallic vitreous • diaphaneity: transparent to translucent • fusibility: 3 • other: commonly decrepitates in a blowpipe flame and loses its color; colors a blowpipe flame red but this can be difficult to see; isotropic; commonly strongly fluoresces blue in long-wavelength UV light; sometimes becomes phosphorescent after heating in a test tube; occasionally triboluminescent □ use: jewelry and decorative arts, including especially small vessels

B. Mohs Hardness > 2.5 to ≤ 5.5				
2. Fusibility > 3				
a. Specific Gravity ≤ 2.70				
white to pale-green	usually various shades of green; rarely brown, yellow, pink, or red	2.6-3.3	2-3	**chlorite, $(Mg,Fe)_3(Si,Al)_4O_{10}(OH)_2 \cdot (Mg,Fe)_3(OH)_6$** [clay and mica groups, a.k.a. clinochloro, chamosite, and green mica] *(monoclinic or triclinic depending on crystallization conditions)* • breakage: cleavage in 1 direction (basal pinacoidal), prominent with easy separation • tenacity: flexible but not elastic • habit: usually in aggregates that are internally foliated and externally massive; and rarely in pseudo-hexagonal tabular crystals • luster: nonmetallic vitreous to pearly or dull • diaphaneity: transparent to translucent • fusibility: 5-5.5 • other: yields water condensate when heated in a test tube; in a blowpipe flame whitens with fusion along the edges; greasy feel □ use: none known, but it is a common rock-forming mineral
white	usually white or various shades of green or yellow, and commonly a mottling of these colors; occasionally gray, brown, or reddish-brown	2.55-2.61	2.5-5	**serpentine group, $Mg_3Si_2O_5(OH)_4$** *(hexagonal, monoclinic or, less often, orthorhombic depending on crystallization conditions)* • <u>varieties</u>: **antigorite** and **lizardite** – granular aggregates, not distinguishable macroscopically, and **chrysotile** [a.k.a. clinochrysotile and asbestos] – fibrous aggregates • breakage: cleavage in 1 (basal pinacoidal) or 2 (prismatic) directions depending on variety, none prominent; also uneven to splintery fracture • tenacity: tough • habit: always in aggregates that are internally compact granular or fibrous (asbestiform) and externally massive • luster: nonmetallic waxy or greasy (granular) to silky (fibrous) • diaphaneity: translucent to pseudo-opaque • fusibility: 5-5.5 • other: yields water condensate when heated in a test tube; commonly fluoresces yellow or white in short- and long-wavelength UV lights □ use: fire-proof cloth for chrysotile; jewelry and decorative arts for serpentinite, a rock consisting mostly of serpentine minerals

white	white to pale-gray; occasionally red	2.60-2.90	3.5-4	**alunite, $KAl_3(SO_4)_2(OH)_6$** [a.k.a. alumite, alum rock, and alumstone] *(hexagonal)* • breakage: cleavage in 1 direction (basal pinacoidal), not prominent; also even, uneven, or conchoidal fracture for crystals and splintery fracture for aggregates • tenacity: brittle • habit: usually in aggregates that are internally granular or fibrous and externally massive or stalactitic; and rarely in rhombohedral or hexagonal tabular crystals • luster: nonmetallic vitreous to pearly (crystals) or dull (aggregates) • diaphaneity: transparent to translucent • fusibility: 7 • other: in a blowpipe flame decrepitates, emits sulfurous fumes, and colors the flame purplish-red but this can be difficult to see; yields a highly acidic water condensate when heated in a test tube; readily dissolves in warm water □ use: roasted to produce potash-alum (see **alum group**)
white	usually light- to dark-green or bluish-green; occasionally colorless, white, pale shades of brown, yellow, or red	2.57-2.61	3.5-4.5	**variscite, $Al(PO_4)\cdot 2H_2O$** [a.k.a. lucinite and utahlite] *(orthorhombic)* • breakage: cleavage in 2 directions (both pinacoidal), none prominent; also uneven to conchoidal or splintery fracture • tenacity: brittle • habit: usually in aggregates that are internally fine granular and externally massive or colloform; and rarely in pseudo-octahedral crystals • luster: nonmetallic vitreous to waxy • diaphaneity: transparent to translucent • fusibility: 6-7 □ use: jewelry and decorative arts

B. Mohs Hardness > 2.5 to ≤ 5.5				
2. Fusibility > 3				
b. Specific Gravity > 2.70				
white to pale-green	usually various shades of green; rarely brown, yellow, pink, or red	2.6-3.3	2-3	**chlorite, $(Mg,Fe)_3(Si,Al)_4O_{10}(OH)_2{\bullet}(Mg,Fe)_3(OH)_6$** [clay and mica groups; a.k.a. clinochlore, chamosite, and green mica] *(monoclinic or triclinic depending on crystallization conditions)* • breakage: cleavage in 1 direction (basal pinacoidal), prominent with easy separation • tenacity: flexible but not elastic • habit: usually in aggregates that are internally foliated and externally massive; and rarely in pseudo-hexagonal tabular crystals • luster: nonmetallic vitreous to pearly or dull • diaphaneity: transparent to translucent • fusibility: 5-5.5 • other: yields water condensate when heated in a test tube; in a blowpipe flame whitens with fusion along the edges; greasy feel □ use: none known, but it is a common rock-forming mineral
white to pale-gray	usually dark-brown or black (weathers to yellow); rarely dark-green for biotite — usually pale-yellow or yellowish-brown; rarely greenish- to reddish-brown for phlogopite	2.70-3.30	2-3	**biotite, $K(Mg,Fe)_3(AlSi_3O_{10})(OH)_2$** to **phlogopite, $KMg_3(AlSi_3O_{10})(OH)_2$** in a solid-solution series [mica group; a.k.a. black mica for biotite, and amber, brown or bronze mica for phologopite] *(monoclinic)* • breakage: cleavage in 1 direction (basal pinacoidal), prominent with easy separation • tenacity: flexible and elastic • habit: commonly in pseudo-hexagonal tabular or short prismatic crystals; and also commonly in aggregates that are internally foliated and externally massive • luster: submetallic to mainly nonmetallic vitreous or pearly • diaphaneity: transparent to translucent • fusibility: 5 • other: yields water condensate when heated in a test tube; immediately whitens in a blowpipe flame with fusion along the edges; biotite is weakly attracted to a neodymium magnet □ use: none known, but these are common rock-forming minerals

white	usually colorless or pale-yellow (weathers to golden yellow); rarely pale-brown, pale-green or pale-red	2.70-2.90	2-3.5	**muscovite, $KAl_2(AlSi_3O_{10})(OH)_2$** [mica group; a.k.a. common, white or potash mica, mirror stone, and Muscovy glass] *(monoclinic)* • breakage: cleavage in 1 direction (basal pinacoidal), prominent with easy separation • tenacity: flexible and elastic • habit: usually in coarse aggregates that are internally foliated and externally massive; and occasionally in pseudo-hexagonal tabular crystals • luster: nonmetallic vitreous to pearly or silky • diaphaneity: transparent to translucent • fusibility: 4-5 • other: whitens in a blowpipe flame and fuses along the edges; yields water condensate when heated in a test tube; thin flakes held up to a bright light commonly exhibit asterism □ use: jewelry and decorative arts; small windows; mirrors; glitter in paint
white	colorless to white or pale-gray, commonly tinged with red, yellow, or brown; rarely pale-green or blue	4.48-4.50	2.5-3.5	**barite, $BaSO_4$** [a.k.a. cawk, barytes, and heavy spar] *(orthorhombic)* • breakage: cleavage in 4 directions (prismatic + two pinacoidal), one prominent; also uneven fracture • tenacity: brittle • habit: usually in tabular or prismatic crystals, commonly in roseiform groups; and occasionally in aggregates that are internally granular and externally massive • luster: nonmetallic vitreous or resinous to pearly • diaphaneity: transparent to pseudo-opaque • fusibility: 3-4 • other: decrepitates in a blowpipe flame and colors it yellowish-green; commonly fluoresces yellowish-white in long-wavelength UV light □ use: ingredient in vitreous glazes

white to pale-gray	usually colorless or white when pure; yellow, brown to reddish-brown, pale-blue, pale-green, pale-purple, gray or black when impure	2.71 pure up to 2.94 impure	3	**calcite, $CaCO_3$** [calcite group] *(hexagonal)* • special varieties: (1) **dog-tooth spar** – combination scalenohedral and prismatic crystal; (2) **nail-head spar** – combination rhombohedral and prismatic crystal; (3) **Iceland spar** – colorless, transparent crystal; (4) **satin spar** – finely fibrous aggregate; (5) **travertine** – fibrous to granular aggregates that are either porous [a.k.a. tufa] or dense [a.k.a. calcite alabaster, oriental alabaster, and cave onyx] • breakage: cleavage in 3 directions (rhombohedral), all prominent • tenacity: brittle • habit: commonly in hexagonal prismatic or tabular crystals, and also in rhombohedral or scalenohedral crystals (commonly in drusy groups); and commonly in aggregates that are internally fibrous to mainly compact to coarse granular and externally massive, colloform, stalactitic, oolitic, or pisolitic • luster: nonmetallic vitreous (crystals) or dull (aggregates) • diaphaneity: transparent to pseudo-opaque • fusibility: 7 • other: colors a blowpipe flame brownish-red but this can be difficult to see; effervesces strongly in cold dilute HCl acid; stained pink or red by Alizarin Red S dye; commonly fluoresces pink, red, or white in short- and long-wavelength, and blue in short-wavelength UV lights; commonly phosphoresces bluish-white or pink in short- and long-wavelength UV lights; sometimes triboluminescent □ use: quicklime in cement, mortar and plaster; jewelry and decorative arts; pottery temper; white paint pigment
white	colorless to light- or dark-gray, occasionally pale-blue or pale-violet (commonly stained by iron oxides)	2.89-3.00	3-3.5	**anhydrite, $CaSO_4$** [a.k.a. cube spar] *(orthorhombic)* • breakage: cleavage in 3 directions (all pinacoidal and mutually perpendicular), one prominent; also uneven to splintery fracture • tenacity: brittle • habit: usually in aggregates that are internally fine granular or fibrous and externally massive; and rarely in slender prismatic or tabular crystals that are in radiated or plumose groups • luster: nonmetallic vitreous to pearly • diaphaneity: transparent to translucent • fusibility: 3-3.5 • other: in a blowpipe flame turns opaque while fusing to a shiny bead, and colors the flame a brownish-red but this can be difficult to see; unlike its close cousin, gypsum, does not dissolve in hot HCl acid □ use: plaster and mortar; white paint pigment

white	usually light- to dark-brown or yellowish-, reddish- or grayish-brown, (darkening with exposure to air)	3.96 (pure) down to 3.5 (impure)	3.5-4	**siderite, $FeCO_3$** [calcite group; a.k.a. chalybite, siderose, iron spar, and sparry, spathose or spathic iron] *(hexagonal)* • breakage: cleavage in 3 directions (rhombohedral), all prominent; also uneven fracture • tenacity: brittle • habit: usually in rhombohedral to, less commonly, hexagonal tabular crystals with curved faces; and occasionally in aggregates that are internally compact to coarse granular and externally massive, botryoidal, oolitic, or earthy and pulverulent • luster: nonmetallic vitreous to pearly • diaphaneity: transparent to translucent, rarely pseudo-opaque • fusibility: 5 • other: decrepitates and blackens in a blowpipe flame; effervesces either weakly or not at all in cold dilute HCl acid when whole but always weakly when powdered; moderately attracted to a neodymium magnet □ use: smelted for iron
white	white to pale-gray; occasionally red	2.60-2.90	3.5-4	**alunite, $KAl_3(SO_4)_2(OH)_6$** [a.k.a. alumite, alum rock, and alumstone] *(hexagonal)* • breakage: cleavage in 1 direction (basal pinacoidal), not prominent; also even, uneven, or conchoidal fracture for crystals and splintery fracture for aggregates • tenacity: brittle • habit: usually in aggregates that are internally granular or fibrous and externally massive or stalactitic; and rarely in rhombohedral or hexagonal tabular crystals • luster: nonmetallic vitreous to pearly (crystals) or dull (aggregates) • diaphaneity: transparent to translucent • fusibility: 7 • other: in a blowpipe flame decrepitates, emits sulfurous fumes, and colors the flame purplish-red but this can be difficult to see; yields a highly acidic water condensate when heated in a test tube; readily dissolves in warm water □ use: roasted to produce potash-alum (see **alum group**)

white	usually colorless or white; occasionally gray or yellow, and rarely blue, green, red, or violet	2.93-2.95	3.5-4	**aragonite**, $CaCO_3$ [aragonite group: a.k.a. Aragon spar] *(orthorhombic)* • gem varieties: **mother-of-pearl** – the strongly iridescent nacre or inner lining of oyster shells; and **pearl** – the faintly iridescent nodules with a pearly luster from inside an oyster (see also Table 5.18) • other biogenic variety: pelecypod (a.k.a. bivalve mollusc) shell (see also Table 5.18) • breakage: cleavage in 1 direction (pinacoidal), not prominent; also sub-conchoidal fracture • tenacity: brittle • habit: commonly in pseudo-hexagonal acicular, short to long prismatic, or tabular crystals that are in radiated groups; and also commonly in aggregates that are internally granular or fibrous and externally colloform, stalactitic, or pisolitic • luster: nonmetallic vitreous • diaphaneity: transparent to translucent • fusibility: 7 • other: decrepitates and whitens in a blowpipe flame and colors it brownish-red but this can be difficult to see; effervesces strongly in cold dilute HCl acid; stained red or pink by Alizarin Red S dye; commonly fluoresces bluish-white or orange in short-wavelength and yellowish-white in long-wavelength UV lights; commonly phosphoresces white or greenish-white in short- and long-wavelength UV lights; sometimes triboluminescent □ use: as mother-of-pearl and pearl – jewelry and decorative arts; as pelecypod shell – beads and quicklime for cement, mortar, and plaster

white	usually colorless, white, gray, or pale-pink; commonly with brownish or yellowish tinges	2.85-2.93	3.5-4	**dolomite, $CaMg(CO_3)_2$** to **ankerite, $CaFe(CO_3)_2$** in a solid-solution series [a.k.a. bitter, brown, magnesian, pearl or bitter spar for dolomite, and cleat spar and ferroan dolomite for ankerite] *(hexagonal)* • breakage: cleavage in 3 directions (rhombohedral), all prominent; also uneven to conchoidal fracture • tenacity: brittle • habit: commonly in rhombohedral crystals with curved faces, and rarely in prismatic or tabular crystals; and also commonly in aggregates that are internally compact to coarse granular and externally massive or stalactitic • luster: nonmetallic vitreous to pearly (crystals) or dull (aggregates) • diaphaneity: transparent to translucent • fusibility: 7 • other: colors a blowpipe flame yellowish-red but this can be difficult to see; effervesces weakly in cold dilute HCl but strongly when powdered; not stained by Alizarin Red S; occasionally triboluminescent □ use: ingredient in vitreous glazes
white	colorless, white, pale-gray, pale-yellow, or brown	3.00-4.38	3.5-4.5	**magnesite, $MgCO_3$** – rarely pure [calcite group; a.k.a. giobertite] *(hexagonal)* • breakage: cleavage in 3 directions (rhombohedral), all prominent • tenacity: brittle • habit: usually in soft (earthy or pulverulent) aggregates that are internally fibrous or compact to coarse granular and externally massive; and rarely in hexagonal prismatic or rhombohedral crystals • luster: nonmetallic vitreous to pearly (crystals) or dull (earthy) • diaphaneity: transparent to translucent • fusibility: 7 • other: weak to mainly no effervescence in cold dilute HCl acid when whole but effervesces strongly when powdered; commonly fluoresces bluish-white in both short- and long-wavelength UV lights □ use: ingredient in vitreous glazes; medicine

white	usually light- to dark-gray or brown; occasionally white, greenish- to brownish-white, green, bluish-green, blue, yellow, or pink	4.30-4.45	4-4.5	**smithsonite, $ZnCO_3$** [calcite group; a.k.a. calamine, szaskaite, zinc spar, and dry-bone ore] *(hexagonal)* • breakage: cleavage in 3 directions (rhombohedral), all prominent; also uneven fracture • tenacity: brittle • habit: usually in aggregates that are internally compact to coarse granular and externally colloform or stalactitic (commonly macroporous); and rarely in small rhombohedral or scalenohedral crystals • luster: nonmetallic vitreous to pearly or waxy • diaphaneity: translucent to pseudo-opaque • fusibility: 7 • other: colors a blowpipe flame bluish-green but this can be difficult to see; weak to mainly no effervescence in cold dilute HCl acid when whole, but weakly (5% HCl) or strongly (10% HCl) effervescent when powdered □ use: smelted for zinc, an ingredient of brass; medicine; metal solder; ingredient in vitreous glazes
white	usually mottled light- and dark-blue; occasionally white, gray, green, grayish-green, pink, yellow, or black	3.53-3.65	4-5.5 parallel to crystal length 6-7 perpendicular to crystal length	**kyanite, Al_2SiO_5** [a.k.a. cyanite, sappare, and disthene] *(triclinic)* • breakage: cleavage in 2 directions (both pinacoidal), one prominent • tenacity: flexible • habit: almost always in tabular to mainly long, bladed crystals that are commonly bent or twisted • luster: nonmetallic vitreous to pearly • diaphaneity: transparent to pseudo-opaque • fusibility: 7 □ use: none known, but it is a common rock-forming mineral

white	usually white or pale-gray; occasionally colorless, light-green, yellowish-brown, or brownish-red	2.86-3.09	4.5-5	**wollastonite, $CaSiO_3$** [pyroxenoid group; a.k.a. tabular spar] *(triclinic)* • breakage: cleavage in 3 directions (all pinacoidal), one prominent; also splintery fracture • tenacity: brittle • habit: usually in aggregates that are internally compact granular to mainly fibrous and externally massive; and rarely in tabular crystals • luster: nonmetallic vitreous to pearly or silky • diaphaneity: translucent • fusibility: 4 • other: commonly fluoresces yellow or orangish-yellow in short-wavelength, and pale-yellow in long-wavelength UV lights; commonly phosphoresces red in short- and long-wavelength UV lights □ use: none known, but it is a common rock-forming mineral
white	usually white or pale-gray; commonly stained by iron oxides (brown to yellowish-brown) or copper oxides (tinges of blue or green)	3.40-3.50	4.5-5	**hemimorphite, $Zn_4(Si_2O_7)(OH)_2{\cdot}H_2O$** [a.k.a. calamine, electric calamine, and galmei] *(orthorhombic)* • breakage: cleavage in 4 directions (two prismatic), two (for one prismatic pair) prominent; also uneven fracture • tenacity: brittle • habit: usually in tabular to bladed prismatic crystals that commonly occur in drusy or radiated groups; and occasionally in aggregates that are internally granular or fibrous and externally massive, mammillary, or stalactitic • luster: nonmetallic vitreous to adamantine or pearly • diaphaneity: transparent to translucent • fusibility: 6 • other: decrepitates in blowpipe flame; commonly fluoresces pale-orange or pale-yellow in short- and long-wavelength UV lights; strongly pyroelectric and piezoelectric □ use: smelted for zinc, an ingredient of brass; medicine; metal solder; ingredient in vitreous glazes

<table>
<tr><td>white</td><td>usually greenish-yellow or bluish-, grayish- to yellowish-green; occasionally pale-pink, orange, brown, brownish-red, or red</td><td>3.15-3.35</td><td>5</td><td>apatite, $Ca_5(PO_4)_3F$ [a.k.a. fluorapatite and tavistockite] (hexagonal)
• breakage: cleavage in 1 direction (basal pinacoidal), not prominent; also uneven to conchoidal fracture
• tenacity: brittle
• habit: usually in hexagonal dipryamidal, prismatic, or tabular crystals with rounded edges; and occasionally in aggregates (commonly referred to as 'collophane') that are internally compact to coarse granular or fibrous and externally massive, botryoidal, oolitic/pisolitic, or stalactitic
• luster: nonmetallic vitreous or greasy to mainly resinous
• diaphaneity: transparent to pseudo-opaque
• fusibility: 5
• other: colors a blowpipe flame yellowish-red but this can be difficult to see; commonly fluoresces yellow, brownish-yellow, or greenish-yellow in short- and long-wavelength UV lights
□ use: jewelry, especially beads</td></tr>
<tr><td colspan="5">C. Mohs Hardness > 5.5</td></tr>
<tr><td colspan="5">1. Fusibility ≤ 3</td></tr>
<tr><td colspan="5">Specific Gravity > 2.70</td></tr>
<tr><td>white</td><td>usually pink to red and commonly with black veins of Mn-oxide; occasionally brown or brownish-red, and rarely green or yellow</td><td>3.40-3.76</td><td>5.5-6.5</td><td>rhodonite, $MnSiO_3$ usually with Ca and Fe impurities [pyroxenoid group; a.k.a. manganese spar and manganolite] (triclinic)
• breakage: cleavage in 3 directions (all pinacoidal), two prominent at a nearly 90° angle; also uneven to conchoidal fracture
• tenacity: tough
• habit: usually in aggregates that are internally fine to coarse granular and externally massive; and rarely in tabular crystals with rounded edges
• luster: nonmetallic vitreous
• diaphaneity: translucent to pseudo-opaque
• fusibility: 2.5-3.5
• other: weakly to moderately attracted to a neodymium magnet
□ use: possibly decorative arts</td></tr>
</table>

white	usually green or white with green patches; sometimes colorless, plain white, blue, greenish-blue, violet, or brown	3.24-3.50	6-7	**jadeite, $NaAlSi_2O_6$** [pyroxene group; a.k.a. jade] *(monoclinic)* • breakage: cleavage in 1 direction (pinacoidal), prominent • tenacity: tough • habit: almost always in aggregates that are internally compact granular and externally massive • luster: nonmetallic vitreous to waxy, greasy, or pearly • diaphaneity: translucent • fusibility: 2.5 □ use: jewelry and decorative arts; tools and weapons
white	usually brown or, less commonly, green or yellow, and rarely blue or red	3.32-3.45	6-7	**vesuvianite, $Ca_{10}(Mg,Fe)_2Al_4(SiO_4)_5(Si_2O_7)_2(OH)_4$** [epidote group; a.k.a. vesuvian and idocrase] *(tetragonal)* • breakage: cleavage in 2 directions (prismatic), none prominent; also uneven fracture • tenacity: brittle • habit: usually in prismatic crystals that are commonly in parallel groups, have longitudinal striations, a square cross-section, and are terminated by a truncated pyramid; and occasionally in aggregates that are internally granular and externally massive • luster: nonmetallic vitreous to resinous • diaphaneity: translucent to pseudo-opaque • fusibility: 3 □ use: none known, but it is a common rock-forming mineral

C. Mohs Hardness > 5.5				
2. Fusibility > 3				
a. Specific Gravity ≤ 2.70				
white	usually colorless or white; sometimes pale shades of gray, brown, red, or multi-colored	2.00-2.50	5-6	**opal, SiO_2•nH_2O** *(mineraloid)* • <u>gem varieties</u>: **precious opal** with a play-of-colors, including **black opal** – flashes of brilliant colors on a black background; **fire opal** – flashes of red, orange and/or yellow on a translucent background; **milky opal** – white and translucent with a hint of opalescence; **hyalite** – colorless and transparent • breakage: no cleavage; conchoidal fracture • tenacity: brittle • habit: always in aggregates that are internally granular and externally massive, botryoidal, stalactitic, or pisolitic; commonly replacing fossil wood • luster: nonmetallic vitreous to waxy or resinous • diaphaneity: transparent to mainly translucent • fusibility: 7 • other: yields water condensate when heated in a test tube; isotropic; usually opalescent; commonly fluoresces yellowish-green in short-wavelength, and pale-yellow or yellowish-green in long-wavelength UV lights; commonly phosphoresces white in short- and long-wavelength UV lights □ use: jewelry and decorative arts
white or pale-green to pale-greenish-blue	light-blue, bluish-green, or green (blue color can degrade over time to green)	2.60-2.91	5-6	**turquoise, $CuAl_6(PO_4)_4(OH)_8$•$4H_2O$** [a.k.a. calaite and Turkey stone] *(triclinic)* • breakage: prominent cleavage in 2 directions (both pinacoidal), one prominent; also even to conchoidal fracture • tenacity: brittle • habit: almost always in aggregates that are internally compact granular and externally massive, colloform, or stalactitic • luster: nonmetallic vitreous to waxy • diaphaneity: translucent to pseudo-opaque • fusibility: 7 • other: yields abundant water condensate when heated in a test tube □ use: jewelry and decorative arts

white to pale-gray	usually white to light-gray; occasionally colorless	2.47-2.50	5.5-6	**leucite, $KAlSi_2O_6$** [feldspathoid group; a.k.a. amphigene, grenatite, and white garnet] *(tetragonal and commonly pseudo-isometric)* • breakage: no cleavage; conchoidal fracture • tenacity: brittle • habit: usually in trapezohedral crystals and less commonly in other multi-faceted, equidimensional crystal forms • luster: nonmetallic vitreous or dull • diaphaneity: translucent to opaque • fusibility: 7 □ use: none known, but it is a common rock-forming mineral
white	colorless, white to light-gray or yellowish-white when crystals, and bluish- to greenish-gray, purplish-brown, or red when aggregates	2.55-2.66	5.5-6	**nepheline, (Na,K)$AlSiO_4$** [feldspathoid group; a.k.a. eleolite and nephelite] *(hexagonal)* • breakage: cleavage in 3 directions (prismatic + basal pinacoidal), none prominent; also conchoidal fracture • tenacity: brittle • habit: almost always in aggregates that are internally compact granular and externally massive; rarely in small, short prismatic crystals with flat tops and bottoms • luster: nonmetallic vitreous (crystals) to greasy (aggregates) • diaphaneity: transparent to mainly translucent or pseudo-opaque • fusibility: 4 □ use: none known, but it is a common rock-forming mineral
white	usually white to pale-gray or pink; rarely colorless, yellow, or green	2.56-2.57	6	**feldspar, potassic – orthoclase, $KAlSi_3O_8$** [feldspar group: a.k.a. orthose] *(monoclinic)* • gem variety: **adularia** [a.k.a. moonstone] – colorless to white, transparent to translucent, and opalescent in blue and white colors • breakage: cleavage in 3 directions (all pinacoidal), one prominent and another nearly so at a ~90° angle • tenacity: brittle • habit: usually in short prismatic crystals; and occasionally in aggregates that are internally fine to coarse granular and externally massive • luster: nonmetallic vitreous to pearly • diaphaneity: transparent to mainly translucent or pseudo-opaque • fusibility: 5 • other: colors a blowpipe flame violet but this can be difficult to see □ use: for adularia, jewelry and decorative arts; none for the common variety, a major rock-forming mineral

white	usually pink to, less often, shades of red; rarely white, pale-yellow, green, or bluish-green	2.54-2.57	6-6.5	**feldspar, potassic – microcline, $KAlSi_3O_8$** [feldspar group] *(triclinic)* • gem variety: **amazonite** [a.k.a. amazon stone] – green to bluish-green • breakage: cleavage in 3 directions (all pinacoidal), one prominent and another nearly so at a ~90° angle • tenacity: brittle • habit: usually in short prismatic crystals with parallel, thin vein-like bands of exsolved albite (i.e. perthite lamellae); and occasionally in aggregates that are internally fine to coarse granular and externally massive • luster: nonmetallic vitreous to pearly • diaphaneity: translucent to pseudo-opaque • fusibility: 5 • other: colors a blowpipe flame violet but this can be difficult to see □ use: for amazonite (and also rarely for the pink variety), jewelry and decorative arts; none for the common variety, a major rock-forming mineral
white	usually white or pale-gray to darker shades of gray with increasing An; rarely colorless, pink, red, reddish-brown, yellow, green, or black	2.62-2.76 increasing with An	6-6.5	**feldspar, plagioclase**, solid-solution series from **$NaAlSi_3O_8$** to **$CaAl_2Si_2O_8$** [feldspar group] *(triclinic)* • mineral varieties: **albite** (An = 0-10), **oligoclase** (An = 10-30), **andesine** (An = 30-50), **labradorite** (An = 50-70), **bytownite** (An = 70-90), and **anorthite** (An = 90-100) where An = (Ca/ [Na + Ca]) x 100 • gem varieties: **labradorite** [a.k.a. spectrolite] – labradorite with well-developed labradorescence; **moonstone** – opalescent albite or oligoclase; and **sunstone** [a.k.a. aventurine] – oligoclase or labradorite with shimmering or sparkling red hematite inclusions • breakage: cleavage in 3 directions (all pinacoidal), one prominent and another nearly so at a ~90° angle • tenacity: brittle • habit: almost always in tabular to mainly bladed (and lath-like) prismatic crystals for albite to oligoclase, but less commonly as crystals for andesine to bytownite and rarely for anorthite, which tends to occur instead as irregular grains • luster: nonmetallic vitreous to pearly • diaphaneity: transparent to pseudo-opaque • fusibility: 4-4.5 • other: parallel striations from polysynthetic twinning is commonly observed on crystal faces and cleavage surfaces □ use: for the gem varieties, jewelry and decorative arts; none for the common varieties, all major rock-forming minerals

white	usually colorless, gray or white, but when impure can be shades of orange, red, brown, yellow, blue, or green	2.50-2.67	6.5-7	**chalcedony** or **chalcedonic quartz**, $SiO_2 \cdot nH_2O$ *(hexagonal)* • gem varieties: **carnelian** (or **cornelian**) - orange, orangish- to brownish-red, or red, and translucent (commonly heat-treated in antiquity to enhance and darken the reddish color); **sard** - orangish-, yellowish- to reddish-brown, or brown, and translucent; **heliotrope** [a.k.a. bloodstone] - green with red spots, and semitranslucent to opaque; **sapphirine** - pale-bluish-gray to grayish-blue, and translucent; **milky chalcedony** - white, and semitransparent to translucent; **agate** - color-banded, translucent chalcedony with sub-varieties **onyx** - linear banding with alternating white or pale-gray and black or dark-gray, **sardonyx** - linear banding with alternating white or pale-gray and red or brownish-red, and **banded agate** - curved, concentric banding with alternating colors as in onyx and sardonyx (especially in geodes); **dendritic** and **moss agates** - plant-like growths of brown or black manganese oxides in white to yellowish-white, translucent chalcedony; **jasper** - green, brown, yellow, red, or combinations of these colors, and pseudo-opaque to opaque; **prase** - dull green, and translucent; **chrysoprase** -yellowish-green, and semi-transparent to translucent; **plasma** - bright-green speckled with white spots, and pseudo-opaque; **tiger's** (or **cat's**) **eye** - chalcedony pseudomorph after the fibrous (asbestiform) chrysotile variety of serpentine, usually in shades of brown and yellow or yellow and blue, and translucent to opaque; **petrified** [a.k.a. silicified or agatized] **wood** - fossil wood where the woody components have been replaced by chalcedony (and minor normal quartz) with a variety of colors that are mostly due to iron oxide impurities, and translucent to opaque • breakage: no cleavage; conchoidal fracture • tenacity: brittle • habit: always in aggregates that are internally compact to fine granular or fibrous and externally massive, colloform, or stalactitic • luster: nonmetallic waxy or dull • diaphaneity: translucent to opaque • fusibility: 7 • other: commonly fluoresces green or pale-yellow in short-wavelength UV light □ use: for gem varieties, jewelry and decorative arts; for common varieties, tools and weapons, hones/whetstones

white	usually colorless or white; when impure can be shades of gray, yellow, brown, red, or violet	2.65	7	**(normal) quartz, SiO_2** *(hexagonal)* • gem varieties: **rock crystal** – colorless, and transparent; **amethyst** – pale-to dark-violet, and transparent; **rose quartz** – pink or pale-red, and transparent to translucent (gradational with amethyst); **citrine** – light-yellow to orange or orangish-brown, and transparent (usually produced in antiquity by heat-treating amethyst; a.k.a. ametrine); **smoky quartz** [a.k.a. cairngorm stone] – yellowish-brown, gray, or nearly black, and transparent to pseudo-opaque; **rutilated quartz** – rock crystal with abundant inclusions of needle-like (acicular) rutile crystals, and transparent; **aventurine** – rock crystal with abundant bright, sparkling scaly red (iron oxide) and green (fuchsite mica) inclusions, and translucent to opaque • other varieties: **milky** [a.k.a. hydrothermal or vein] **quartz** – white, and translucent; **ferruginous quartz** – brownish-red to reddish-brown, and transparent to translucent • breakage: no cleavage; conchoidal fracture • tenacity: brittle • habit: commonly in hexagonal prismatic crystals with pyramidal terminations and parallel transverse striations (perpendicular to the long axis) that are often found in drusy groups, especially in geodes; and also commonly in aggregates that are internally compact to coarse granular and externally massive • luster: nonmetallic vitreous to occasionally greasy • diaphaneity: mainly transparent to translucent, but occasionally pseudo-opaque or opaque • fusibility: 7 • other: rose quartz commonly exhibits pleochroism with alternating brownish-red and light-pink; strongly piezoelectric, pyroelectric, and readily develops frictional electricity □ use: for gem varieties, jewelry and decorative arts; for common varieties, tools and weapons, ingredient in glass and vitreous glazes

white	usually various shades of blue or violet; occasionally colorless, gray, yellow, or brown	2.53-2.78	7-7.5	**cordierite, $(Mg,Fe)_2 Al_4Si_5AlO_{18} \cdot nH_2O$** *(orthorhombic)* gem varieties: **iolite** [a.k.a. dichroite] – violet and blue-hued, transparent to translucent, and pleochroic; and **bloodshot iolite** – iridescent blue, violet and red, and translucent • breakage: cleavage in 2 directions (both pinacoidal), none prominent; also sub-conchoidal fracture • tenacity: brittle • habit: commonly in short pseudo-hexagonal crystals; and also commonly in aggregates that are internally granular and externally massive • luster: nonmetallic vitreous to resinous • diaphaneity: transparent to mainly translucent • fusibility: 5-5.5 • other: loses transparency in a blowpipe flame; commonly exhibits pleochroism with alternating colorless to pale-yellow, blue, and dark-blue to violet; readily develops frictional electricity □ use: jewelry and seals
white	usually colorless, white, pale-green, or pale-yellow; occasionally blue, pink, and other shades of green and yellow	2.65-2.95	7.5-8	**beryl, $Be_3Al_2Si_6O_{18}$** *(hexagonal)* gem varieties, all transparent to semi-transparent: **emerald** – dark-green to dark-bluish-green; **aquamarine** – pale-blue to greenish-blue; **heliodor** – pale- to vivid-yellow; **morganite** – pale- to vivid-pink; and **goshenite** – colorless (note that the translucent color varieties are sometimes referred to by the gemstone names) • breakage: no cleavage; uneven to conchoidal fracture • tenacity: brittle • habit: always in long, hexagonal prismatic crystals that commonly have flat terminations • luster: nonmetallic vitreous to resinous • diaphaneity: transparent to translucent, rarely pseudo-opaque • fusibility: 5-5.5 • other: commonly exhibits pleochroism with alternating green and bluish-green for emerald, and light-red and reddish-purple for morganite □ use: for gem varieties, jewelry and the decorative arts; for common variety, possibly an abrasive

C. Mohs Hardness > 5.5				
2. Fusibility > 3				
b. Specific Gravity > 2.70				
white	light- to dark-green or brownish-green for actinolite — colorless, white, or light-gray for tremolite (but pink or pale-violet when Mn-bearing)	2.98-3.48	5-6	**actinolite, $Ca_2(Mg,Fe)_5Si_8O_{22}(OH)_2$** [a.k.a. ferro-actinolite] to **tremolite, $Ca_2Mg_5Si_8O_{22}(OH)_2$** in a solid-solution series [amphibole group; a.k.a. **nephrite** or **jade** when a compact massive aggregate] *(monoclinic)* • breakage: cleavage in 4 directions (prismatic + two pinacoidal), two (prismatic) prominent at ~56° and ~124° angles • tenacity: brittle except nephrite which is tough • habit: usually in acicular to long bladed prismatic crystals that are commonly in radiated groups; and occasionally in aggregates that are internally fibrous (asbestiform) or compact to coarse granular and externally massive • luster: nonmetallic vitreous to silky • diaphaneity: transparent to translucent • fusibility: 4 • other: actinolite is weakly attracted to a neodymium magnet, and tremolite is not attracted unless iron-bearing and then only weakly □ use: for nephrite/jade – tools and weapons; jewelry and decorative arts
white to pale-gray or pale-greenish-gray	usually greenish-black; occasionally dark-green, dark-brown, or black	3.00-3.59	5-6	**hornblende, $(Ca,Na)_{2\text{-}3}(Al,Fe,Mg)_5Si_6(Al,Si)_2O_{22}(OH)_2$** [amphibole group] *(monoclinic)* • breakage: cleavage in 4 directions (prismatic + two pinacoidal), two (prismatic) prominent at ~56° and ~124° angles • tenacity: brittle • habit: commonly either in long, sometimes acicular, prismatic crystals or irregular grains; and also commonly in aggregates that are internally granular to mainly fibrous and externally massive • luster: nonmetallic vitreous to silky • diaphaneity: translucent to pseudo-opaque • fusibility: 4 • other: fuses, sometimes with intumescence, to a magnetic globule; colors a blowpipe flame a conspicuous vivid yellow; weakly attracted to a neodymium magnet □ use: none known, but it is a common rock-forming mineral

white	usually brown; occasionally colorless, gray, green, yellow, or black	3.40-3.60	5-6	**titanite, $CaTiO(SiO_4)$** [a.k.a. grothite and sphene] *(monoclinic)* • breakage: cleavage in 2 directions (prismatic), none prominent; also sub-conchoidal fracture • tenacity: brittle • habit: usually in wedge-shaped dipyramidal crystals; and occasionally in aggregates that are internally granular or foliated and externally massive • luster: nonmetallic adamantine to mainly resinous • diaphaneity: transparent to translucent • fusibility: 4 • other: in blowpipe flame fuses with intumescence to a yellow or black globule □ use: none known, but it is a common rock-forming mineral
white or pale-green to pale-greenish-blue	light-blue, bluish-green, or green (blue color can degrade over time to green)	2.60-2.91	5-6	**turquoise, $CuAl_6(PO_4)_4(OH)_8{\cdot}4H_2O$** [a.k.a. calaite and Turkey stone] *(triclinic)* • breakage: prominent cleavage in 2 directions (both pinacoidal), one prominent; also even to conchoidal fracture • tenacity: brittle • habit: almost always in aggregates that are internally compact granular and externally massive, colloform, or stalactitic • luster: nonmetallic vitreous to waxy • diaphaneity: translucent to pseudo-opaque • fusibility: 7 • other: yields abundant water condensate when heated in a test tube □ use: jewelry and decorative arts
white	usually in shades of brown, especially dark-yellowish-brown; occasionally white, gray, or green	2.85-3.57	5.5-6	**anthophyllite, $Mg_7Si_8O_{22}(OH)_2$** [amphibole group; a.k.a. bidalotite] *(orthorhombic)* • breakage: cleavage in 4 directions (prismatic + two pinacoidal), two (prismatic) prominent • tenacity: brittle • habit: usually in parallel or radiated groups of acicular crystals; occasionally in aggregates that are internally fibrous (asbestiform) and externally massive • luster: nonmetallic vitreous to pearly (crystals) or silky (aggregates) • diaphaneity: transparent to mainly translucent or pseudo-opaque • fusibility: 5-6 • other: becomes magnetic when fused □ use: none known, but it is a common rock-forming mineral

white to pale-green	dark-green or greenish- to brownish-black	3.19-3.56	5.5-6.5	**augite, $(Ca,Na)(Mg,Fe,Al)(Si,Al)_2O_6$** [pyroxene group: a.k.a. basaltine] *(monoclinic)* • breakage: cleavage in 4 directions (two prismatic), two (for one prismatic pair) prominent at a nearly 90° angle • tenacity: brittle • habit: commonly in short prismatic crystals; and also commonly in aggregates that are internally granular and externally massive or rarely stalactitic • luster: nonmetallic vitreous to resinous • diaphaneity: translucent to pseudo-opaque • fusibility: 4-4.5 □ use: none known, but it is a common rock-forming mineral
white to pale-gray	usually white or light-green; occasionally dark-green	3.22-3.30	5.5-6.5	**diopside, $CaMgSi_2O_6$** [pyroxene group: a.k.a. malacolite] *(monoclinic)* • breakage: cleavage in 4 directions (two prismatic), none prominent; also uneven to conchoidal fracture • tenacity: brittle • habit: usually in short, prismatic crystals with square or octagonal cross-sections; and occasionally in aggregates that are internally granular and externally massive to stalactitic • luster: nonmetallic vitreous • diaphaneity: transparent to translucent • fusibility: 4 □ use: none known, but it is a common rock-forming mineral
white	usually pink to red and commonly with black veins of Mn-oxide; occasionally brown or brownish-red, and rarely green or yellow	3.40-3.76	5.5-6.5	**rhodonite, $MnSiO_3$** usually with Ca and Fe impurities [pyroxenoid group; a.k.a. manganese spar and manganolite] *(triclinic)* • breakage: cleavage in 3 directions (all pinacoidal), two prominent at a nearly 90° angle; also uneven to conchoidal fracture • tenacity: tough • habit: usually in aggregates that are internally fine to coarse granular and externally massive; and rarely in tabular crystals with rounded edges • luster: nonmetallic vitreous • diaphaneity: translucent to pseudo-opaque • fusibility: 2.5-3.5 • other: weakly to moderately attracted to a neodymium magnet □ use: possibly decorative arts

white	usually white or pale-gray to darker shades of gray with increasing An; rarely colorless, pink, red, reddish-brown, yellow, green, or black	2.62-2.76 increasing with An	6-6.5	**feldspar, plagioclase**, solid-solution series from $\mathbf{NaAlSi_3O_8}$ to $\mathbf{CaAl_2Si_2O_8}$ [feldspar group] *(triclinic)* • mineral varieties: **albite** (An = 0-10), **oligoclase** (An = 10-30), **andesine** (An = 30-50), **labradorite** (An = 50-70), **bytownite** (An = 70-90) and **anorthite** (An = 90-100) where An = (Ca/ [Na + Ca]) x 100 • gem varieties: **labradorite** [a.k.a. spectrolite] – labradorite with well-developed labradorescence; **moonstone** – opalescent albite or oligoclase; and **sunstone** [a.k.a. aventurine] – oligoclase or labradorite with shimmering or sparkling red hematite inclusions • breakage: cleavage in 3 directions (all pinacoidal), one prominent and another nearly so at a ~90° angle • tenacity: brittle • habit: almost always in tabular to mainly bladed (and lath-like) prismatic crystals for albite to oligoclase, but less commonly as crystals for andesine to bytownite and rarely for anorthite, which tends to occur instead as irregular grains • luster: nonmetallic vitreous to pearly • diaphaneity: transparent to pseudo-opaque • fusibility: 4-4.5 • other: parallel striations from polysynthetic twinning is commonly observed on crystal faces and cleavage surfaces □ use: for the gem varieties, jewelry and decorative arts; none for the common varieties, all major rock-forming minerals
white to pale-gray	yellowish-green, brownish-green, or blackish-green; rarely colorless, pale-yellow, or gray	3.21-3.49	6-7	**epidote,** $\mathbf{Ca_2(Al,Fe)Al_2O(SiO_4)(Si_2O_7)(OH)}$ [epidote group; a.k.a. arendalite, delphinite, thallite, and pistacite] *(monoclinic)* • breakage: cleavage in 2 directions (both pinacoidal), one prominent; also uneven fracture • tenacity: brittle • habit: commonly in prismatic crystals that are in radiated groups; and also commonly in aggregates that are internally fine to coarse granular or fibrous and externally massive • luster: nonmetallic vitreous • diaphaneity: transparent to mainly translucent or pseudo-opaque • fusibility: 3-4 • other: fuses with intumescence to a weakly magnetic globule; when heated in a test tube yields a little water; sometimes weakly attracted to a neodymium magnet □ use: jewelry and decorative arts

white	colorless, pale-gray, yellow, yellowish-brown, green, pink, or purple	3.15-3.36	6-7	**zoisite, $Ca_2Al_3Si_3O_{12}(OH)$** [epidote group] *(orthorhombic)* • gem variety: **tanzanite** – blue to purplish-blue and transparent • breakage: cleavage in 1 direction, prominent; also uneven to conchoidal fracture • tenacity: brittle • habit: commonly in prismatic crystals; and also commonly in aggregates that are internally compact granular and externally massive or stalactitic • luster: nonmetallic vitreous to pearly • diaphaneity: transparent to translucent • fusibility: 3-3.5 • other: fuses with intumescence in a blowpipe flame; tanzanite commonly exhibits pleochroism with alternating blue, purple, and greenish-yellow □ use: for tanzanite, possibly jewelry; none for the common variety, a rock-forming mineral

white or a pale-shade of the whole mineral color	mostly various shades of red, purple, green, brown, or yellow; occasionally white, gray, or black	3.13-3.59 hydro-grossular 3.58 pyrope 3.59 grossular 3.86 andradite 3.83-3.90 uvarovite 4.19 spessartine 4.32 almandine	6-7 hydro-grossular 7-7.5 all others	**garnet group, $\alpha_3\beta_2Si_3O_{12}$** in a solid-solution series where α = Ca, Fe, Mg and/or Mn, and β = Al, Cr and/or Fe *(isometric)* • varieties: **almandine** [a.k.a. almandite], $Fe_3Al_2Si_3O_{12}$ - red, purplish-red, purple, reddish-brown, or brownish-black; **pyrope**, $Mg_3Al_2Si_3O_{12}$ - red to purplish-red or nearly black; **spessartine**, $Mn_3Al_2Si_3O_{12}$ - orange to reddish-orange, dark-red to brownish-red, brown, or black; **grossular** [a.k.a. grossularite and hessonite] $Ca_3Al_2Si_3O_{12}$ - greenish-white, pale- to dark-green, less often colorless, yellow, pink, reddish-orange, or brown; **andradite**, $Ca_3Fe_2Si_3O_{12}$ - yellowish-green, brownish-green, brown, greenish-brown, yellow, or black; **uvarovite**, $Ca_3Cr_2Si_3O_{12}$ - dark- to vivid-green; and **hydrogrossular**, $Ca_3Al_2Si_2O_8(SiO_4)_{1-m}(OH)_{4m}$ where m varies between 0 and 1 - green to bluish-green, white, gray, or pink • breakage: no cleavage; conchoidal to uneven fracture • tenacity: brittle • habit: usually in crystals that are dodecahedral and sometimes trapezohedral or tetrahexahedral; and rarely in aggregates that are internally granular and externally massive • luster: nonmetallic vitreous to resinous • diaphaneity: transparent to translucent (pyrope), translucent only (almandine, grossular, spessartine, and uvarovite), or pseudo-opaque to opaque (andradite and hydrogrossular) • fusibility: 3-4, except uvarovite which is 7 • other: all varieties exhibit isotropy when transparent or semi-transparent; iron-rich almandine and andradite are weakly to mainly moderately or strongly attracted to a neodymium magnet, and manganese-rich spessartine is only weakly attracted □ use: for the gem varieties, jewelry and decorative arts; for the common varieties, an abrasive (except hydrogrossular)

white	usually dark-green or greenish-gray to greenish-black; occasionally bluish-gray	3.40-3.8	6.5	**chloritoid, $(Fe,Mg)_2Al_4O_2(SiO_4)_2(OH)_4$** *(usually monoclinic but occasionally triclinic depending on crystallization conditions)* • breakage: cleavage in 2 directions (two pinacoidal), one prominent • tenacity: brittle • habit: usually in aggregates that are internally foliated in radial splays and externally massive; and rarely in tabular, pseudo-hexagonal crystals • luster: nonmetallic vitreous to pearly • diaphaneity: translucent • fusibility: 6-7 • other: yields water when heated in a test tube; commonly becomes weakly magnetic after heating in a blowpipe flame □ use: none known, but it is a common rock-forming mineral
white to pale-gray	usually green to yellowish-green (all varieties); occasionally white to yellow (forsterite) or brown to black (fayalite)	3.22-4.39 (increasing with Fe content)	6.5-7	**olivine, $(Mg,Fe)_2SiO_4$** [a.k.a. chrysolite and olivinoid], a solid-solution series from **forsterite, Mg_2SiO_4** to **fayalite, Fe_2SiO_4** *(orthorhombic)* • gem variety: **peridot** – transparent and light-green to yellowish-green • breakage: cleavage in 2 directions (two pinacoidal), none prominent; also conchoidal fracture • tenacity: brittle • habit: usually in aggregates that are internally granular and externally massive; and occasionally in tabular to short, somewhat flattened prismatic crystals, and rarely in acicular crystals • luster: nonmetallic vitreous • diaphaneity: transparent to translucent • fusibility: 7 • other: fayalite is weakly attracted to a neodymium magnet □ use: for peridot, jewelry and decorative arts; none for the common variety, a major rock-forming mineral

white	usually white, pink, red, purplish-red, or reddish-brown; rarely yellow or green (can be dark-gray to nearly black from common carbonaceous impurities)	3.14-3.20	6.5-7.5	**andalusite, Al_2SiO_5** [a.k.a. macle] *(orthorhombic)* • breakage: cleavage in 3 directions (prismatic + pinacoidal), none prominent; also uneven to sub-conchoidal fracture • tenacity: brittle • habit: almost always in prismatic crystals with square cross-sections that are commonly in radiated groups • luster: nonmetallic vitreous to dull • diaphaneity: transparent to mainly translucent to opaque • fusibility: 7 • other: crystal cross-sections usually exhibit a black cross (due to carbonaceous impurities) within a white square; commonly pleochroic with alternating brownish-green and dark-red □ use: none known, but it is a common rock-forming mineral
white	usually colorless, white to light-gray, or pale-green; occasionally bluish- to grayish-green, or brown	3.23-3.27	6.5-7.5	**sillimanite, Al_2SiO_5** [a.k.a. fibrolite] *(orthorhombic)* • breakage: cleavage in 1 direction (pinacoidal), prominent; also uneven fracture • tenacity: brittle • habit: usually in long, slender to acicular, four-sided prismatic crystals with a diamond-shaped cross-section that commonly occur in parallel or radiated groups; and occasionally in aggregates that are internally fibrous and externally massive • luster: nonmetallic vitreous to silky • diaphaneity: transparent to translucent • fusibility: 7 □ use: none known, but it is a common rock-forming mineral

white	usually various shades of blue or violet; occasionally colorless, gray, yellow, or brown	2.53-2.78	7-7.5	**cordierite, $(Mg,Fe)_2 Al_4Si_5AlO_{18}•nH_2O$** *(orthorhombic)* <u>gem varieties</u>: **iolite** [a.k.a. dichroite] – violet and blue-hued, transparent to translucent, and pleochroic; and **bloodshot iolite** – iridescent blue, violet and red, and translucent • breakage: cleavage in 2 directions (both pinacoidal), none prominent; also sub-conchoidal fracture • tenacity: brittle • habit: commonly in short pseudo-hexagonal crystals; and also commonly in aggregates that are internally granular and externally massive • luster: nonmetallic vitreous to resinous • diaphaneity: transparent to mainly translucent • fusibility: 5-5.5 • other: loses transparency in a blowpipe flame; commonly exhibits pleochroism with alternating colorless to pale-yellow, blue, and dark-blue to violet; readily develops frictional electricity □ use: jewelry and seals
white or a pale-shade of the whole mineral color	usually black or bluish-black; occasionally green, blue, brown, pink, or with multi-color banding perpendicular to crystal length	2.90-3.25 increasing with Fe content	7-7.5	**tourmaline** group *(hexagonal)* • <u>common varieties</u>: **dravite, $NaMg_3Al_6Si_6O_{18}(BO_3)_3(OH)_3(OH,F)$** – usually brown; **schorl, $NaFe_3^{2+}Al_6Si_6O_{18}(BO_3)_3(OH)_3(OH,F)$** – most abundant variety, usually black; **elbaite, $Na(Li_{1.5},Al_{1.5})Al_6Si_6O_{18}(BO_3)_3(OH)_3(OH,F)$** – usually shades of green, blue, yellow, pink, or red (this is also the principal gem variety) • breakage: cleavage in 2 directions (prismatic), none prominent; also uneven to conchoidal fracture • tenacity: brittle • habit: almost always in long, acicular to slender, prismatic crystals, with rounded triangular cross-sections and parallel longitudinal striations on faces, that are commonly in radiated groups; and rarely in aggregates that are internally granular and externally massive • luster: nonmetallic vitreous to resinous • diaphaneity: transparent to pseudo-opaque • fusibility: 3-5 • other: fuses with intumescence in a blowpipe flame; commonly pleochroic with alternating lighter and darker shades of the whole mineral color (when transparent or semi-transparent); schorl is moderately to mainly weakly attracted to a neodymium magnet; commonly piezoelectric and thermoluminescent □ use: jewelry and decorative arts

white to pale-gray	dark-brown, reddish- to yellowish-brown, or brownish-black	3.65-3.85	7-7.5	**staurolite, $Fe_{3-4}Al_{18}Si_8O_{48}H_{2-4}$** [a.k.a. cross stone, fairy stone, grenatite, and staurotide] *(monoclinic to pseudo-orthorhombic)* • breakage: cleavage in 1 direction (pinacoidal), not prominent; also uneven to conchoidal fracture • tenacity: brittle • habit: almost always in prismatic crystals that commonly occur in interpenetrating cruciform (cross-shaped) twins • luster: nonmetallic vitreous to resinous • diaphaneity: translucent to pseudo-opaque • fusibility: 7 • other: weakly to moderately attracted to a neodymium magnet □ use: none known, but it is a common rock-forming mineral; however, it has striking twinned crystals that may have attracted attention in antiquity
white	usually brown or reddish-brown; occasionally colorless, gray, pale-yellow, green, violet, or red	4.60-4.70	7.5	**zircon, $ZrSiO_4$** *(tetragonal)* • gem varieties, all transparent: **hyacinth** – orange to reddish-brown or red; **jargon** – colorless to pale-yellow; **zirconite** – gray or brown; **starlite** – blue but produced by heat-treating common zircon • breakage: cleavage in 2 directions (prismatic), none prominent; also conchoidal fracture • tenacity: brittle • habit: almost always in prismatic crystals with pyramidal terminations • luster: nonmetallic adamantine to vitreous • diaphaneity: transparent to translucent • fusibility: 7 • other: light colored varieties become colorless or white in a blowpipe flame; hyacinth commonly exhibits pleochroism with alternating reddish-purple and reddish-brown; commonly fluoresces orange, orangish-yellow or pale-yellow in short-wavelength, and pale-yellow in long-wavelength UV lights □ use: possibly jewelry

white	usually colorless, white, pale-green, or pale-yellow; occasionally blue, pink, and other shades of green and yellow	2.65-2.95	7.5-8	**beryl, $Be_3Al_2Si_6O_{18}$** *(hexagonal)* gem varieties, all transparent to semi-transparent: **emerald** – dark-green to dark-bluish-green; **aquamarine** – pale-blue to greenish-blue; **heliodor** – pale- to vivid-yellow; **morganite** – pale- to vivid-pink; and **goshenite** – colorless (note that the translucent color varieties are sometimes referred to by the gemstone names) • breakage: no cleavage; uneven to conchoidal fracture • tenacity: brittle • habit: always in long, hexagonal prismatic crystals that commonly have flat terminations • luster: nonmetallic vitreous to resinous • diaphaneity: transparent to translucent, rarely pseudo-opaque • fusibility: 5-5.5 • other: commonly exhibits pleochroism with alternating green and bluish-green for emerald, and light-red and reddish-purple for morganite □ use: for gem varieties, jewelry and the decorative arts; for the common variety, possibly an abrasive
white to pale-gray, pale-green, or pale-brown	colorless, white, red, yellow, brown, blue, green, or black	3.55-4.10	7.5-8	**spinel, $MgAl_2O_4$** [a.k.a. spinelite] *(isometric)* • gem variety: **ruby spinel** – red and transparent • breakage: no cleavage; conchoidal fracture • tenacity: brittle • habit: commonly in octahedral, less often cubical or dodecahedral, crystals; and also commonly in aggregates that are internally compact to coarse granular and externally massive • luster: nonmetallic vitreous • diaphaneity: transparent to mainly translucent or pseudo-opaque • fusibility: 7 • other: isotropic; commonly fluoresces red in long-wavelength UV light □ use: jewelry; possibly inscribing tools

white	usually colorless or white; occasionally light-gray, pale-blue, yellow, yellowish-brown, green, pink, or red	3.40-3.60	8	**topaz, $Al_2SiO_4(F,OH)_2$** *(orthorhombic)* • gem variety: **precious topaz** – orange to yellow, and transparent • breakage: cleavage in 1 direction (pinacoidal), prominent; also uneven to sub-conchoidal fracture • tenacity: brittle • habit: usually in prismatic crystals with pyramidal terminations and longitudinally striated faces; and occasionally in aggregates that are internally fine to coarse granular and externally massive or stalactitic • luster: nonmetallic vitreous • diaphaneity: transparent to translucent • fusibility: 7 • other: precious topaz commonly exhibits pleochroism with alternating brown to brownish-yellow and yellowish-brown to orangish-yellow; readily develops frictional electricity □ use: possibly for jewelry
white	various shades of green	3.65-3.75	8.5	**chrysoberyl, $BeAl_2O_4$** [a.k.a. chrysopal and gold beryl] *(orthorhombic)* gem varieties: **alexandrite** – transparent, dark-green in reflected daylight but usually red in reflected artificial light, and red to purple in transmitted light; **cat's eye** or **cymophane** – chatoyant and translucent • breakage: cleavage in 4 directions (prismatic + two pinacoidal), none prominent; also uneven to conchoidal fracture • tenacity: brittle • habit: almost always in prismatic or tabular crystals, often with longitudinal striations and commonly in six-sided stellated groups of twinned crystals • luster: nonmetallic vitreous • diaphaneity: transparent to translucent • fusibility: 7 • other: often red in transmitted light when green in reflected light; commonly exhibits pleochroism with alternating dark-red, orange, and green; alexandrite commonly fluoresces red in short- and long-wavelength UV lights □ use: possibly in jewelry and decorative arts; possibly an abrasive

white	colorless, white, gray, yellow, yellowish-brown, pink, red, orange, blue, purple, and occasionally color-zoned	3.98-4.02	9	**corundum, Al_2O_3** [a.k.a. corindon, and adamantine or diamond spar] *(hexagonal)* gem varieties, all transparent: **blue sapphire** – blue; **ruby** – pink to red; other 'sapphires', include **oriental amethyst** – purple, **oriental emerald** – green, and **oriental topaz** – yellow • breakage: no cleavage, but has prominent parting in 1 direction that resembles basal pinacoidal cleavage; also uneven to conchoidal fracture • tenacity: brittle • habit: usually in hexagonal tabular to short prismatic and rarely dipyramidal crystals having somewhat rounded edges with prismatic crystals thicker in the middle (i.e. barrel-shaped) and having transverse striations on the faces; and occasionally in aggregates that are internally fine to coarse granular and externally massive (called **emery** when impure with iron oxides) • luster: nonmetallic adamantine to vitreous (crystals) or dull (aggregates) • diaphaneity: transparent to translucent • fusibility: 7 • other: blue sapphire often exhibits asterism; commonly exhibits pleochroism with alternating dark-violet and light-greenish-blue for blue sapphire, purplish-red and orangish-red for ruby, violet and orange for oriental amethyst, and green and yellow for oriental emerald; commonly fluoresces red or orangish-red in short- and long-wavelength UV lights; readily develops frictional electricity □ use: for blue sapphire and rarely for ruby, jewelry; for common corundum and especially emery, an abrasive
white	usually colorless or pale-yellow; occasionally white, pale-pink or red, orange or pale-green; rarely blue or black	3.52	10	**diamond, C** [a.k.a. native carbon] *(isometric)* • breakage: cleavage in 4 directions (octahedral), all prominent; also conchoidal fracture • tenacity: brittle • habit: almost always in octahedral and, less often, cubical, tetrahedral, or dodecahedral crystals, commonly with curved faces • luster: nonmetallic adamantine (cut) to greasy (uncut) • diaphaneity: transparent to translucent • fusibility: 7 • other: isotropic; commonly fluoresces blue in short- and long-wavelength UV lights; commonly phosphoresces bluish-white in long-wavelength UV light; displays a brilliant play-of-colors in reflected light when cut □ use: an abrasive; jewelry and decorative arts, but only beginning in the medieval period

Table 3.7: Fluorescent minerals in Tables 3.4-3.6 and 5.18.

Fluorescence[1]					Mineral
General Color	Specific Color				
	Short-Wavelength (SW)	Occur.[2]	Long-Wavelength (LW)	Occur.[2]	
reddish or pinkish	red, purplish-red or pink	—	red, purplish-red or pink	—	albite plagioclase
			red	—	andalusite
	red or orangish-red	—			andesine plagioclase
			red	—	anhydrite
	red	—			ankerite
	red	—			anorthite plagioclase
	violetish-pink	—	pale-pink or violetish-pink	—	anthophyllite
	pink or violetish-pink	—	orangish-red, pink, or violetish-pink	—	apatite
	red, orangish-red, violetish-red, pink, or violetish-pink	—	orangish-red or violetish-red, pink, or violetish-pink	—	aragonite
	pink	—			barite
	pink	—	pale-red or pink	—	beryl
	red or pink	█	red or pink	█	calcite
	red or pink	█	red or pink	█	calomel
	orangish-red	■	orangish-red	■	
	red	█	red	█	alexandrite chrysoberyl
	red, violetish-red, pink, or violetish-pink	■			cordierite
	red or pink	■	red or pink	█	corundum
			violetish-red or violetish-pink	—	
			red or pink	—	diamond
	red, orangish-red, violetish-red, pink, or violetish-pink	—	red, orangish-red, violetish-red, pink, or violetish-pink	—	dolomite
	red	■			epidote
	red	—	red or violetish-pink	—	fluorite
	red	■	red	■	grossular garnet
			violetish-red or pink	—	
	red or pink	—	orangish-red or pink	—	halite
	red	■	orangish-red	■	haüyne

[1] Data from Gleason (1960: 26-103 and 108-113), the www.fluomin.org.uk database of 'luminescent minerals', and *www.uvminerals.org* (the 'Fluorescent Mineral Society').

[2] Frequency of occurrence codes: █ = often present, ■ = occasionally present, and — = rarely present. Empty cells for a given mineral indicate that no fluorescence has been reported for that wavelength.

Fluorescence[1]					Mineral
General Color	Specific Color				
	Short-Wavelength (SW)	Occur.[2]	Long-Wavelength (LW)	Occur.[2]	
reddish or pinkish			red or pink	—	haüyne
			red	—	jadeite
	red or pink	■	red or pink	■	kyanite
	red	—			labradorite plagioclase
	violetish-pink	—			lazurite/lapis lazuli
			red	■	morganite beryl
	red or violetish-red	■			microline feldspar
	red or orangish-red	▮	orangish-red	▮	nepheline
	red	■			oligoclase plagioclase
	red	—			orthoclase
			pink or violetish-pink	—	pearl
	red	—			quartz
	red	■			rhodonite
	violetish-pink	—	violetish-pink	—	sepiolite
	red	—			sillimanite
	red or violetish-red	—	red	—	smithsonite
			orangish-red	■	spessartine garnet
	red	—	red	▮	spinel
	orangish-red	■	red or pink	■	tremolite
			red	■	uvarovite garnet
			orangish-red	—	vesuvianite
			violetish-red	—	wollastonite
brownish			pale-brown	—	anorthite plagioclase
	yellowish-brown	■	yellowish-brown	■	apatite
	brown	—	brown	—	aragonite
orangish	orange	—			albite plagioclase
			orange	—	alunite
			yellowish-orange	▮	anglesite
			orange	—	ankerite
	orange or brownish-orange	■	orange or brownish-orange	■	apatite
	orange	▮			aragonite
	orange	—	orange	—	barite
	dark-orange or brownish-orange	—	pale-orange	—	beryl
	orange	—	orange	—	calcite (if weathered)

Fluorescence[1]					Mineral
General Color	Specific Color				
	Short-Wavelength (SW)	Occur.[2]	Long-Wavelength (LW)	Occur.[2]	
orangish			orange	—	calomel
	brownish-orange	—			chalcedony
	orange	—			corundum
			brownish-orange or dark-orange	—	diamond
	brownish-orange	—	brownish-orange	—	fluorite
	orange or brownish-orange	—			grossular garnet
			orange	—	haüyne
	pale-orange	▮	pale-orange	▮	hemimorphite
	orange	■			hydrogrossular garnet
	orange	—	orange or brownish-orange	—	lazurite/lapis lazuli
	orange	■			mimetite
			orange	—	nepheline
			orange	—	quartz
	orange	■	orange	▮	sphalerite
	brownish-orange	■			titanite
			orange	—	topaz
			orange	■	tremolite
	orange brownish-orange	■ —	orange or brownish-orange	—	vesuvianite
	reddish-orange orange	■ —	reddish-orange	■	wollastonite
	orange	▮			zircon
yellowish	pale-yellow or greenish-yellow	—	pale-yellow or greenish-yellow	—	amber
	pale-yellow	▮	yellow or orangish-yellow	▮	anglesite
	orangish-yellow	—			actinolite
	pale-yellow	—			antigorite serpentine
	yellow, brownish-yellow, greenish-yellow, or orangish-yellow	▮	yellow, brownish-yellow, or greenish-yellow	▮	apatite
	orangish-yellow	—			augite
	pale-yellow or orangish-yellow	—	pale-yellow or orangish-yellow	■	barite
	pale-yellow or orangish-yellow	■	pale-yellow or orangish-yellow	■	beryl
	yellow	—	yellow	—	calcite

Fluorescence[1]					Mineral
General Color	Specific Color				
	Short-Wavelength (SW)	Occur.[2]	Long-Wavelength (LW)	Occur.[2]	
yellowish	yellow	—			cassiterite
	yellow	■	yellow	▮	cerussite
	pale-yellow	▮	pale-yellow	■	chalcedony
			orangish-yellow	—	
	greenish-yellow	—			common chrysoberyl
	yellow	—	yellow	—	corundum
			pale-yellow or greenish-yellow	—	diamond
	orangish-yellow	—			dravite tourmaline
	pale-yellow	—	pale-yellow	—	fluorite
	pale-yellow	—	pale-yellow	—	forsterite olivine
	pale-yellow	▮	pale-yellow	▮	gypsum
	bluish-yellow, brownish-yellow, or greenish-yellow	—	bluish-yellow, brownish-yellow, greenish-yellow, or orangish-yellow	—	
	pale-yellow	▮	pale-yellow	▮	hemimorphite
			orangish-yellow	—	natron
			yellow	■	oligoclase plagioclase
	pale-yellow, greenish-yellow, or orangish-yellow	—	pale-yellow	▮	opal
			greenish-yellow	—	
	yellow	■	yellow, greenish-yellow, or orangish-yellow	—	quartz
	greenish-yellow or orangish-yellow	—			
	pale-yellow, greenish-yellow, or orangish-yellow	—	pale-yellow, greenish-yellow, or orangish-yellow	—	sphalerite
	yellow	▮	yellow	▮	serpentine group
	pale-yellow	—	pale-yellow	—	talc
	pale-yellow	■	yellow	■	topaz
			orangish-yellow	—	
	yellow	—	pale-yellow	—	tremolite
	yellow	■			tourmaline group
	yellow or orangish-yellow	—	orangish-yellow	■	vesuvianite
	yellow or orangish-yellow	▮	pale-yellow	▮	wollastonite
			orangish-yellow	—	
	pale-yellow or orangish-yellow	▮	pale-yellow	▮	zircon

Fluorescence[1]					Mineral
General Color	Specific Color				
	Short-Wavelength (SW)	Occur.[2]	Long-Wavelength (LW)	Occur.[2]	
yellowish	brownish-yellow	—	brownish-yellow	—	zircon
			orangish-yellow	■	zoisite
greenish	green	■	green	■	adamite
	yellowish-green	—			
	green	—			albite plagioclase
	yellowish-green	—	yellowish-green	—	amber
	dark-green or yellowish-green	—			andalusite
	green or yellowish-green	—	green or yellowish-green	—	apatite
	green or yellowish-green	—	yellowish-green	—	aragonite
	pale-green	■			beryl
	green	▮			chalcedony
			green	—	chlorargyrite
	yellowish-green	■	yellowish-green	■	common chrysoberyl
	brownish-green	■			dravite tourmaline
	pale-green, bluish-green, or yellowish-green	—	pale-green or yellowish-green	—	fluorite
	yellowish-green	—			gypsum
	green	—			hemimorphite
	green	—	green	—	jadeite
			green	■	microcline
	yellowish-green	▮	yellowish-green	▮	opal
	green or bluish-green	—	green or bluish-green	—	
	green	■			sillimanite
	green	—			smithsonite
	green	■	green	■	variscite
bluish			blue	—	albite plagioclase
	blue or grayish-blue	■	blue or grayish-blue	■	amber
	pale-blue	—			anthophyllite
	blue	—			antigorite serpentine
	blue	—	violetish-blue	—	apatite
			blue	—	aragonite
	deep blue	—	deep blue	—	amethyst quartz
	blue	▮	blue	—	calcite
			blue	—	chlorargyrite

Fluorescence[1]					Mineral
General Color	Specific Color				
	Short-Wavelength (SW)	Occur.[2]	Long-Wavelength (LW)	Occur.[2]	
bluish	blue	▮	blue	▮	diamond
	blue or greenish-blue	—	blue	—	diopside
			blue	■	elbaite tourmaline
	blue	■			epsomite
	blue	■	blue	▮	fluorite
	violetish-blue	—	violetish-blue	—	
			blue	—	huntite
	blue	■	blue	—	lazurite/lapis lazuli
			blue	—	magnesite
	blue	■	blue	■	pearl
	blue	—			smithsonite
	blue	—			tremolite
	blue	—	blue	—	tourmaline
violetish			violet	—	andesine plagioclase
	violet	■	violet	■	apatite
	violet	—	violet	—	fluorite
			violet	—	labradorite plagioclase
	violet	—			smithsonite
whitish or shades of off-white	greenish-white	■	greenish-white	■	adamite
	white or yellowish-white	—	white or yellowish-white	—	albite plagioclase
	white or yellowish-white	■	white or yellowish-white	■	alunite
	yellowish-white	▮	bluish-white	▮	amber
	white, bluish-white, or greenish-white	—	white, greenish-white, or yellowish-white	—	
	yellowish-white	—			andesine plagioclase
	white or yellowish-white	▮			anglesite
	white	—	yellowish-white	—	anorthite plagioclase
	white	—			anthophyllite
	bluish-white, greenish-white, or yellowish-white	—			antigorite serpentine
	bluish-white, greenish-white, pinkish-white, or yellowish-white	—	bluish-white, greenish-white, pinkish-white, or yellowish-white	—	apatite
	bluish-white	▮	yellowish-white	▮	aragonite
	white, greenish-white, pinkish-white, or yellowish-white	—	white, greenish-white, bluish-white, pinkish-white, or yellowish-white	—	

Fluorescence[1]					Mineral
General Color	Specific Color				
	Short-Wavelength (SW)	Occur.[2]	Long-Wavelength (LW)	Occur.[2]	
whitish or shades of off-white	yellowish-white	■	yellowish-white	■	barite
	white, bluish-white, or greenish-white	—	bluish-white or greenish-white	—	
	bluish-white	—			beryl
	yellowish-white	—	yellowish-white	—	borax
	white	▮	white	▮	calcite
	bluish-white or yellowish-white	—	yellowish-white	▮	cerussite
			bluish-white	—	
	white or yellowish-white	—	greenish-white	—	chalcedony
			greenish-white	—	common chrysoberyl
	white	▮	white	▮	coral
			bluish-white, greenish-white, pinkish-white, or yellowish-white	—	diamond
	bluish-white	—			diopside
	yellowish-white	■	yellowish-white	■	dolomite
	bluish-white, greenish-white, or pinkish-white	—	white, bluish-white, greenish-white, or pinkish-white	—	
	bluish-white	■	bluish-white	—	elbaite tourmaline
	yellowish-white	—			
			yellowish-white	■	epsomite
	white, bluish-white, greenish-white, or yellowish-white	—	white, bluish-white, greenish-white, or yellowish-white	—	fluorite
	yellowish-white	■	yellowish-white	■	forsterite olivine
	yellowish-white	—	bluish-white	—	grossular garnet
	bluish-white	▮	yellowish-white	▮	gypsum
	white, greenish-white, pinkish-white, or yellowish-white	—	white, bluish-white, or greenish-white	—	
	white, bluish-white, or yellowish-white	—	white	—	hemimorphite
	yellowish-white	▮	yellowish-white	▮	huntite
	bluish-white	—			
	bluish-white	▮	bluish-white	▮	ivory
			white	■	jadeite
	white or yellowish-white	■	white or yellowish-white	■	kaolinite

Fluorescence[1]					Mineral
General Color	Specific Color				
	Short-Wavelength (SW)	Occur.[2]	Long-Wavelength (LW)	Occur.[2]	
whitish or shades of off-white	white or yellowish-white	—	white or yellowish-white	—	kyanite
			yellowish-white	—	lazurite/lapis lazuli
	bluish-white	■	yellowish-white	■	magnesite
	white	—	white, bluish-white, or greenish-white	—	magnesite
	bluish-white	—			microcline
			yellowish white	■	mimetite
	yellowish-white	■	yellowish-white	■	muscovite
	white or yellowish-white	■	yellowish-white	■	natron
	yellowish-white	—	yellowish-white	—	nepheline
	greenish-white or yellowish-white	—	greenish-white or yellowish-white	—	opal
			yellowish-white	—	orthoclase
	yellowish-white	■	yellowish-white	■	palygorskite
			white	■	potash-alum
	bluish-white	■	bluish-white	■	pearl
	bluish-white	■	yellowish-white	—	pearl
	yellowish-white	■	white or bluish-white	—	quartz
	bluish-white	—	white or bluish-white	—	quartz
	bluish-white or yellowish-white	■	bluish-white or yellowish-white	■	sepiolite
			yellowish-white	■	sillimanite
	white	■			smectite (montmorillonite)
	greenish-white	■	greenish-white	■	smithsonite
	greenish-white	■	white or yellowish-white	—	smithsonite
	white	—	white or yellowish-white	■	serpentine group
	dull white or yellowish-white	■	yellowish-white	■	talc
	bluish-white or greenish-white	—	white, bluish-white, or greenish-white	—	talc
	yellowish-white	■	yellowish-white	■	topaz
	greenish-white	—	greenish-white	—	topaz
	bluish-white or yellowish-white	—	bluish-white, greenish-white, or yellowish-white	—	tremolite
	bluish-white	—	bluish-white	—	tourmaline group
	white, bluish-white or greenish-white	—	white, bluish-white, or greenish-white	—	variscite

Fluorescence[1]						Mineral
General Color	Specific Color					
	Short-Wavelength (SW)	Occur.[2]	Long-Wavelength (LW)	Occur.[2]		
whitish or shades of off-white	bluish-white or yellowish-white	—	yellowish-white	—		wollastonite
	greenish-white or yellowish-white	—				zircon

Table 3.8: Phosphorescent minerals in Tables 3.4-3.6 and 5.18.

Phosphorescence[1]					Mineral
General Color	Specific Color				
	Short-Wavelength (SW)	Occur.[2]	Long-Wavelength (LW)	Occur.[2]	
reddish or pinkish	pink	■	pink	■	morganite beryl
	red (strong)	—			diamond
	pink	—	pink	—	diopside
	pink	█	pink	█	calcite
	pink	—	pink	—	dolomite
	red	—			tremolite
	red	█	red	█	wollastonite
brownish	brownish	—			albite plagioclase
orangish	yellowish-orange	■	yellowish-orange	■	apatite
	orange	■	orange	■	sphalerite
	orange (strong)	■			vesuvianite
	orange (strong)	■			wollastonite
yellowish	yellow	■	yellow	■	adamite
			pale-yellow	—	chalcedony
	greenish-yellow (strong)	—			opal
	pale-yellow	■	pale-yellow	■	sphalerite
greenish	green	■	green	■	adamite
	greenish	█	greenish	█	borax
bluish	blue	█	blue	█	borax
	blue	■	blue	■	calcite
	pale-blue	—	pale-blue	—	cerussite
	blue	■	blue	■	diamond
	blue	■	blue	■	fluorite
	blue	■	blue	■	lazurite/lapis lazuli
	blue	■			natron
violetish	violet	■	violet	■	fluorite
whitish or off-white			yellowish-white	■	amber
	white or greenish-white (strong)	█	white or greenish-white (strong)	█	aragonite

[1] Data from Gleason (1960: 26-103 and 108-113), Sinkankas (1972: 342), and the *www.fluomin.org.uk* database of 'luminescent minerals'.

[2] Frequency of occurrence codes: █ = often present, ■ = occasionally present, and — = rarely present. Empty cells for a given mineral indicate that no phosphorescence has been reported for that wavelength.

Phosphorescence[1]					Mineral
General Color	Specific Color				
	Short-Wavelength (SW)	Occur.[2]	Long-Wavelength (LW)	Occur.[2]	
whitish or off-white	greenish-white or yellowish-white	■	greenish-white or yellowish-white	■	barite
	bluish-white	—	bluish-white	—	goshenite beryl
	bluish-white (strong)	▮	bluish-white (strong)	▮	calcite
	white or greenish-white	—	white	—	chalcedony
			bluish-white (strong)	▮	diamond
	greenish-white	—	greenish-white (strong)	—	fluorite
	white or bluish-white (strong)	■	white or yellowish-white (strong)	■	gypsum
	white (weak)	■	white (weak)	■	hemimorphite
	greenish-white	■	greenish-white	■	kyanite
			yellowish-white (weak)	■	lazurite/lapis lazuli
	white	—	white	—	magnesite
	white	▮	white	▮	opal
	white (weak)	■	white (weak)	■	quartz
	greenish-white	—	greenish-white	—	smithsonite
	yellowish-white	—	yellowish-white	—	topaz
	bluish-white	■	bluish-white	■	zircon

Table 3.9: Specific gravities of archaeologically significant, geologically common, and economically important minerals.

Specific Gravity	Mineral[1]
1.00-2.60	**palygorskite**
1.46	mirabilite
1.46-1.50	sassolite
1.48	**natron**
1.60	carnallite
1.68	**epsomite**
1.7	aluminite
1.72	**borax**
1.75	**potash-alum**
1.89	melanterite
1.90-2.30	**opal**
1.91	kernite
1.93-2.40	**chrysocolla**
1.95	morenosite
1.96	ulexite
1.98	goslarite
1.99	gaylussite
1.99	sylvite
2.00-2.20	**sepiolite**, clay group
2.00-2.65	bauxite
2.00-3.00	**montmorillonite**, clay group
2.05-2.20	chabazite
2.07	**sulfur, native**
2.09-2.23	**graphite**
2.10-2.20	stilbite
2.11	**niter**
2.11	trona
2.12-2.15	mordenite
2.14-2.17	clinoptilolite
2.15-2.20	heulandite

Specific Gravity	Mineral[1]
2.16-2.17	**halite**
2.20	phillipsite
2.20-2.26	natrolite
2.20-2.80	garnierite
2.23-2.47	thomsonite
2.24-2.29	analcime
2.25	thermonatrite
2.25-2.29	scolecite
2.26	mesolite
2.27	laumontite
2.27	tridymite
2.27-2.29	nitratine
2.27-2.33	sodalite
2.29	**chalcanthite**
2.30	vermiculite, clay group
2.30-2.37	**gypsum**
2.30-2.40	apophyllite, clay group
2.30-2.40	nosean
2.32-2.65	cancrinite – vishnevite series
2.33	cristobalite
2.36	wavellite
2.38-2.42	gibbsite
2.38-2.45	**lazurite**
2.39	brucite
2.40-2.95	**glauconite**, mica group
2.41-2.42	petalite
2.41-2.47	harmotome
2.42	colemanite
2.44-2.50	**haüyne**
2.47-2.50	leucite
2.50-2.60	**antigorite**, serpentine group
2.50-2.60	**chalcedony**
2.50-2.62	marialite, scapolite group

[1] The 354 minerals included in this table are those in Tables 3.4-3.6 plus all the others in Berry *et al.* (1983), Gribble (1988), Klein and Dutrow (2007), and Deer *et al.* (2013). Minerals that were either definitely or possibly used in antiquity are in bold.

Specific Gravity	Mineral[1]	Specific Gravity	Mineral[1]
2.53-2.78	**cordierite**	2.75-2.85	glauberite
≤ 2.5	**chrysotile**, serpentine group	2.76	anorthite, plagioclase feldspar group
2.55	**lizardite**, serpentine group	2.77-2.88	**muscovite**, mica group
2.55-2.63	**orthoclase**, potassic feldspar group	2.78	polyhalite
2.56	**microcline**, potassic feldspar group	2.80-2.90	lepidolite, mica group
2.56	sanidine, feldspar group	2.80-3.00	datolite
2.56-2.66	nepheline, feldspathoid group	2.84-2.86	**dolomite**
2.57	kieserite	2.84-2.90	pectolite
2.57-2.61	**variscite**	2.85	paragonite, mica group
2.57-2.65	zaratite	2.85-3.57	anthophyllite
2.58-2.82	hyalophane, feldspar group	2.86-3.09	wollastonite
2.59-2.62	kalsilite	2.90-2.97	prehnite
2.60	**talc**	2.90-3.00	**anhydrite**
2.60-2.62	**albite**, plagioclase feldspar group	2.90-3.02	zinnwaldite
2.60-2.80	**turquoise**	2.90-3.10	elbaite, tourmaline group
2.60-2.90	alunite	2.90-3.26	**jarosite**
2.60-2.90	collophane apatite	2.91-3.10	boracite
2.60-2.90	**illite**, clay group	2.92	coesite
2.60-2.90	stilpnomelane	2.93-3.10	ankerite
2.60-3.30	chlorite, clay and mica groups	2.94	åkermanite
2.61-2.68	**kaolinite**, clay group	2.94-2.95	**aragonite**
2.63-2.66	**oligoclase**, plagioclase feldspar group	2.95-3.05	melilite
2.65	**quartz** (normal)	2.96-2.98	cryolite
2.65-2.90	pyrophyllite	2.96-3.00	phenakite
2.66	thénardite	2.97-3.02	danburite
2.66-2.92	**beryl**	2.97-3.45	richterite - ferrorichterite series
2.67-2.69	vivianite	2.98-3.02	**magnesite**
2.69	andesine, plagioclase feldspar group	2.99	ferrimolybdite
2.70	**huntite**	2.99-3.08	margarite
2.70-3.30	biotite - phlogopite series, mica group	2.99-3.48	actinolite (as **nephrite**)
2.70-4.30	**limonite**	3.00-3.10	clintonite
2.71	**calcite**	3.00-3.20	eckermannite
2.71	**labradorite**, plagioclase feldspar group	3.00-3.20	**tyrolite**
2.74	bytownite, plagioclase feldspar group	3.02-3.05	böehmite
2.74-2.78	meionite, scapolite group	3.02-3.59	hornblende group

Specific Gravity	Mineral[1]
3.03-3.18	dravite, tourmaline group
3.03-3.23	spodumene
3.03-3.27	monticellite
3.04-3.11	amblygonite
3.05-3.12	lawsonite
3.05-3.15	**glaucophane**
3.05-3.20	autunite
3.06	erythrite
3.07	annabergite
3.07	gehlenite
3.08	**cuprorivaite**
3.10-3.35	**apatite**
3.10-3.39	celsian, feldspar group
3.10-3.60	cummingtonite – grunerite series
3.12-3.24	lazulite
3.14-3.16	andalusite
3.15-3.37	**zoisite**
3.15-3.50	**riebeckite**
3.16-3.25	pumpellyite
3.16-3.26	chondrodite
3.16-3.43	omphacite
3.17-3.35	clinohumite
3.17-3.46	pigeonite
3.18	**fluorite**
3.18	**natrojarosite**
3.18	norbergite
3.18-3.22	schorl, tourmaline group
3.19-3.56	augite – ferroaugite series
3.20-3.30	kaersutite – ferrokaersutite series
3.20-3.50	diaspore
3.20-3.50	katophorite-magnesiocatophorite series
3.20-4.00	hypersthene
3.21-3.38	clinozoisite
3.21-3.41	dumortierite
3.21-3.96	enstatite
3.22	torbernite
3.22-3.28	**forsterite** olivine
3.22-3.38	**dioptase**
3.22-3.56	diopside – hedenbergite series
3.23-3.27	sillimanite
3.24-3.43	**jadeite**
3.27	scorodite
3.27-3.35	axinite group
3.27-3.54	johannsenite
3.30-3.50	arfvedsonite
3.30-3.50	idocrase
3.32-3.34	vesuvianite
3.38-3.49	**epidote**
3.38-3.61	piemontite
3.40	**veszelyite**
3.40-4.20	allanite group
3.42-3.58	triphylite
3.46-3.80	chloritoid
3.48	**hemimorphite**
3.48-3.50	sinhalite
3.48-3.60	**titanite**
3.49	**orpiment**
3.49-3.57	**topaz**
3.50-3.53	**diamond**
3.50-4.00	hydrozincite
3.52	**pararealgar**
3.52-3.70	metatorbernite
3.53-3.65	kyanite
3.55	**spinel**
3.55-3.60	aegirine
3.56	**realgar**
3.56-3.68	periclase
3.57-3.76	**rhodonite**
3.57-4.35	tyuyamunite
3.58	**pyrope**, garnet group
3.59	grossular, garnet group

Specific Gravity	Mineral[1]
3.60	kosmochlor
3.60-4.00	ferrosilite
3.60-4.05	**malachite**
3.66	plumbojarosite
3.66-3.71	barytocalcite
3.67-3.71	alstonite
3.70	rhodochrosite
3.70-4.70	**romanèchite**
3.72-3.74	**paratacamite**
3.74-3.78	strontianite
3.74-3.80	**atacamite**
3.74-3.83	**staurolite**
3.75	**chrysoberyl**
3.77	**azurite**
3.77	**clinoatacamite**
3.77-3.83	uvarovite, garnet group
3.80-3.90	andradite, garnet group
3.82-3.97	anatase
3.88-3.90	**antlerite**
3.89-4.19	willemite
3.93-3.96	**aurichalcite**
3.95-4.04	alabandite
3.95-4.84	perovskite
3.96	ferrihydrite
3.96	**siderite**
3.96-3.98	celestine
3.97	**brochantite**
3.98-4.02	**corundum**
3.99-4.05	ilvaite
4.00-4.10	wurtzite
4.08-4.18	brookite
4.09	lepidocrocite
4.10	**sphalerite**
4.10-4.30	**chalcopyrite**
4.10-4.30	magnesiochromite
4.19	**spessartine**, garnet group
4.23	**rutile**
4.27-4.30	**goethite**
4.29	witherite
4.29-4.34	manganite
4.30-4.50	**stannite**
4.31-4.48	pyrochlore group
4.32	**almandine**, garnet group
4.32-4.48	**adamite**
4.33-4.39	pseudobrookite
4.35	stishovite
4.38	**clinoclase**
4.39	**fayalite**, olivine group
4.40-4.50	**enargite**
4.40-5.85	xenotime
4.42-4.44	**smithsonite**
4.46	**olivenite**
4.48-4.50	**barite**
4.50-5.11	**tennantite-tetrahedrite** series
4.56-4.65	magnesioferrite
4.60-4.70	**zircon**
4.60-4.76	**covellite**
4.60-4.90	pyrrhotite
4.60-5.00	pentlandite
4.62	gahnite
4.62-4.73	molybdenite
4.63	**stibnite**
4.70	carnotite
4.70-4.79	**ilmenite**
4.72-4.83	braunite
4.80-4.90	greenockite
4.83-4.85	hausmannite
4.87	jacobsite
4.88	maghemite
4.88	zippeite
4.89	**marcasite**
4.95	**hollandite**

Specific Gravity	Mineral[1]
4.95-5.02	**pyrite**
5.00-5.30	monazite group
5.04-5.08	**pyrolusite**
5.06-5.09	**bornite**
5.09	**chromite**
5.15-6.57	columbite
5.20	**magnetite**
5.26	**hematite**
5.30-5.50	**millerite**
5.34	franklinite
5.35	linarite
5.50-5.80	**chalcocite**
5.55	**digenite**
5.56	**chlorargyrite**
5.57	proustite
5.63	jamesonite
5.63-5.78	arsenic, native
5.64-5.68	zincite
5.82	pyrargryite
5.83-5.86	**bournonite**
5.88-5.92	tellurite
5.90-6.07	**arsenopyrite**
5.97-6.02	crocoite
6.00-6.23	boulangerite
6.10	polybasite
6.10	scheelite
6.12-6.15	phosgenite
6.14	**cuprite**
6.20-6.23	freieslebenite
6.26	stephanite
6.33	cobaltite
6.37-6.39	**anglesite**
6.45	**tenorite**
6.5	cervantite
6.5	nickelskutterudite
6.50	skutterudite
6.50-7.50	wulfenite
6.53-6.57	**cerussite**
6.55	leadhillite
6.57-8.00	tantalite
6.61-6.71	antimony, native
6.70	thorite
6.78	bismuthinite
6.88	vanadinite
6.98-7.02	**cassiterite**
7.00-8.00	**iron, meteoritic**
7.04	pyromorphite
7.12-7.18	hübnerite, wolframite group
7.15	**calomel**
7.19-7.36	**acanthite**
7.24	**mimetite**
7.30	tetradymite
7.30-7.87	iron, native (or telluric)
7.35-7.49	nagyágite
7.50-7.60	**galena**
7.58	ferberite, wolframite group
7.59-8.23	breithauptite
7.78	nickeline
7.90	**domeykite**
8.16	sylvanite
8.18	**cinnabar**
8.24-8.45	hessite
8.70-9.40	petzite
8.94	**copper, native**
9.10-9.40	calaverite
9.56	**plattnerite**
9.70-9.83	bismuth, native
10.50	**silver, native**
10.58	sperrylite
10.63-10.95	uraninite

Specific Gravity	Mineral[1]
13.60	**mercury, native**
19.3	**gold, native**
21.46	**platinum, native**

Table 3.10: Mohs hardness of minerals in Tables 3.4-3.6.

Mohs Hardness[1]		Mineral[2]
1-2	0	mercury, native
	1	talc
	1-1.5	natron
	1-1.5	pararealgar
	1-2	aurichalcite
	1-2	calomel
	1-2	graphite
	1-2	huntite
	1-2	illite, clay group
	1.5-2	covellite
	1.5-2	orpiment
	1.5-2	realgar
	1.5-2	tyrolite
	1.5-2	vivianite
	1.5-2.5	**montmorillonite**, clay group
	1.5-2.5	sulfur, native
2-3	1.5-2.5	**montmorillonite**, clay group
	1.5-2.5	sulfur, native
	2	glauconite, mica group
	2	gypsum
	2	niter
	2	stibnite
	2-2.5	acanthite
	2-2.5	borax
	2-2.5	cinnabar
	2-2.5	epsomite

Mohs Hardness[1]		Mineral[2]
2-3	2-2.5	halite
	2-2.5	kaolinite, clay group
	2-2.5	palygorskite
	2-2.5	potash-alum
	2-2.5	sepiolite, clay group
	2-3	biotite – phlogopite series, mica group
	2-3	chlorargyrite
	2-3	chlorite, clay and mica groups
	2-3.5	muscovite, mica group
	2-4	chrysocolla
	2-6	pyrolusite
	2.5	chalcanthite
	2.5	galena
	2.5-3	anglesite
	2.5-3	bournonite
	2.5-3	chalcocite
	2.5-3	clinoclase
	2.5-3	copper, native
	2.5-3	digenite
	2.5-3	gold, native
	2.5-3	silver, native
	2.5-3.5	barite
	2.5-3.5	jarosite
	2.5-3.5	natrojarosite
	2.5-5	antigorite, serpentine group
	2.5-5	chrysotile, serpentine group
	2.5-5	lizardite, serpentine group

[1] Minerals are arranged according to their minimum hardness and those with the same hardness are listed alphabetically. Some minerals, because of their hardness range, will appear in two hardness groups.

[2] The minerals included are those in Tables 3.4-3.6.

Mohs Hardness[1]		Mineral[2]
3-4	2-4	chrysocolla
	2-6	pyrolusite
	2.5	chalcanthite
	2.5-3.5	barite
	2.5-3.5	jarosite
	2.5-3.5	natrojarosite
	2.5-5	antigorite, serpentine group
	2.5-5	chrysotile, serpentine group
	2.5-5	lizardite, serpentine group
	3	bornite
	3	calcite
	3	clinoatacamite
	3	enargite
	3	olivenite
	3	paratacamite
	3-3.5	anhydrite
	3-3.5	atacamite
	3-3.5	cerussite
	3-3.5	domeykite
	3-3.5	millerite
	3-4	tenorite
	3-4.5	tennantite-tetrahedrite series
	3.5	adamite
	3.5-4	alunite
	3.5-4	ankerite
	3.5-4	antlerite
	3.5-4	aragonite
	3.5-4	azurite
	3.5-4	brochantite
	3.5-4	cuprite
	3.5-4	dolomite
	3.5-4	malachite
	3.5-4	mimetite

Mohs Hardness[1]		Mineral[2]
3-4	3.5-4	siderite
	3.5-4	sphalerite
	3.5-4	veszelyite
	3.5-4.5	chalcopyrite
	3.5-4.5	magnesite
	3.5-4.5	pyrrhotite
	3.5-4.5	variscite
4-5	2-6	pyrolusite
	2.5-5	antigorite, serpentine group
	2.5-5	chrysotile, serpentine group
	2.5-5	lizardite, serpentine group
	3-4.5	tennantite-tetrahedrite series
	3.5-4.5	chalcopyrite
	3.5-4.5	magnesite
	3.5-4.5	pyrrhotite
	3.5-4.5	variscite
	4	fluorite
	4	stannite
	4-4.5	platinum, native
	4-4.5	smithsonite
	4-5.5	limonite
	4-6	hollandite
	4-7	kyanite
	4.5	iron, meteoritic
	4.5-5	hemimorphite
	4.5-5	wollastonite
5-6	4-5.5	limonite
	4-6	hollandite
	4-7	kyanite
	5	apatite
	5	cuprorivaite
	5	dioptase
	5-5.5	goethite

Mohs Hardness[1]		Mineral[2]
	5-5.5	lazurite
	5-6	actinolite (as nephrite)
	5-6	glaucophane
	5-6	hornblende group
	5-6	ilmenite
	5-6	opal
	5-6	riebeckite
	5-6	romanèchite
	5-6	titanite
	5-6	tremolite
	5-6	turquoise
5-6	5-6.5	hematite
	5.5	plattnerite
	5.5-6	anthophyllite
	5.5-6	arsenopyrite
	5.5-6	chromite
	5.5-6	haüyne
	5.5-6	leucite
	5.5-6	nepheline
	5.5-6.5	augite – ferroaugite series
	5.5-6.5	diopside – hedenbergite series
	5.5-6.5	magnetite
	5.5-6.5	rhodonite
	4-7	kyanite
	5-6.5	hematite
	5.5-6.5	augite – ferroaugite series
	5.5-6.5	diopside – hedenbergite series
6-7	5.5-6.5	magnetite
	5.5-6.5	rhodonite
	6	orthoclase, potassic feldspar group
	6-6.5	albite, plagioclase feldspar group
	6-6.5	andesine, plagioclase feldspar group
	6-6.5	anorthite, plagioclase feldspar group

Mohs Hardness[1]		Mineral[2]
	6-6.5	bytownite, plagioclase feldspar group
	6-6.5	labradorite, plagioclase feldspar group
	6-6.5	marcasite
	6-6.5	microcline, potassic feldspar group
	6-6.5	oligoclase, plagioclase feldspar group
	6-6.5	pyrite
	6-6.5	rutile
	6-7	cassiterite
	6-7	epidote
6-7	6-7	hydrogrossular, garnet group
	6-7	jadeite
	6-7	vesuvianite
	6-7	zoisite
	6.5	chloritoid
	6.5-7	chalcedony
	6.5-7	fayalite olivine
	6.5-7	forsterite olivine
	6.5-7.5	andalusite
	6.5-7.5	sillimanite
	6.5-7.5	andalusite
	6.5-7.5	sillimanite
	7	quartz (normal)
	7-7.5	almandine, garnet group
	7-7.5	andradite, garnet group
	7-7.5	cordierite
7-8	7-7.5	dravite, tourmaline group
	7-7.5	elbaite, tourmaline group
	7-7.5	grossular, garnet group
	7-7.5	pyrope, garnet group
	7-7.5	schorl, tourmaline group
	7-7.5	spessartine, garnet group
	7-7.5	staurolite
	7-7.5	uvarovite, garnet group

Mohs Hardness[1]		Mineral[2]
7-8	7.5	zircon
	7.5-8	beryl
	7.5-8	spinel

Mohs Hardness[1]		Mineral[2]
8-9	8	topaz
	8.5	chrysoberyl
9-10	9	corundum
10	10	diamond

Table 3.11: Fusibility of minerals in Tables 3.4-3.6.

Fusibility[1]		Mineral[2]
0	0	calomel
	0	cinnabar
	0	mercury, native
1	1	bournonite
	1	chlorargyrite
	1	enargite
	1	epsomite
	1	galena
	1	natron
	1	niter
	1	orpiment
	1	potash-alum
	1	realgar
	1	stibnite
	1	sulfur, native
1-2	1-1.5	borax
	1.5	acanthite
	1.5	cerussite
	1.5	halite
	1.5	mimetite
	1.5	stannite
	1.5	tennantite-tetrahedrite series
	1.5-2	millerite
2	2	arsenopyrite
	2	domeykite
	2	plattnerite
	2	silver, native
2-3	2.5	anglesite
	2.5	bornite

Fusibility[1]		Mineral[2]
2-3	2.5	gold, native
	2.5	jadeite
	2.5	marcasite
	2.5	pyrite
	2.5-3	pyrrhotite
	2.5-3.5	rhodonite
	2-2.5	chalcocite
	2-2.5	chalcopyrite
	2-2.5	clinoclase
	2-2.5	covellite
	2-2.5	olivenite
	2-2.5	tyrolite
	2-2.5	vivianite
3	3	adamite
	3	azurite
	3	chalcanthite
	3	copper, native
	3	cuprite
	3	fluorite
	3	malachite
	3	tenorite
	3	vesuvianite
3-4	3.5	brochantite
	3.5	haüyne
	3.5	lazurite
	3-3.5	anhydrite
	3-3.5	glaucophane
	3-3.5	gypsum
	3-3.5	riebeckite
	3-3.5	zoisite
	3-4	almandine, garnet group
	3-4	andradite, garnet group
	3-4	atacamite

[1] Minerals are arranged according to their minimum fusibility and those with the same fusibility are listed alphabetically. Some minerals, because of their fusibility range, will appear in two fusibility groups.
[2] The minerals included are those in Tables 3.4-3.6.

Fusibility[1]		Mineral[2]
3-4	3-4	barite
	3-4	epidote
	3-4	grossular, garnet group
	3-4	hydrogrossular, garnet group
	3-4	pyrope, garnet group
	3-4	spessartine, garnet group
	≤ 4	veszelyite
	3-5	dravite, tourmaline group
	3-5	elbaite, tourmaline group
	3-5	schorl, tourmaline group
	3-5?	antlerite
4	4	actinolite (as nephrite)
	4	diopside – hedenbergite series
	4	hornblende group
	4	nepheline
	4	titanite
	4	tremolite
	4	wollastonite
4-5	3-5	dravite, tourmaline group
	3-5	elbaite, tourmaline group
	3-5	schorl, tourmaline group
	3-5?	antlerite
	4.5	jarosite
	4.5	natrojarosite
	4-4.5	albite, plagioclase feldspar group
	4-4.5	andesine, plagioclase feldspar group
	4-4.5	anorthite, plagioclase feldspar group
	4-4.5	augite – ferroaugite series
	4-4.5	bytownite, plagioclase feldspar group
	4-4.5	labradorite, plagioclase feldspar group
	4-5	muscovite, mica group

Fusibility[1]		Mineral[2]
5	5	apatite
	5	biotite – phlogopite series, mica group
	5	microcline, potassic feldspar group
	5	orthoclase, potassic feldspar group
	5	sepiolite, clay group
	5	siderite
5-6	5.5	goethite
	5-5.5	antigorite, serpentine group
	5-5.5	beryl
	5-5.5	chlorite, clay and mica groups
	5-5.5	chrysotile, serpentine group
	5-5.5	cordierite
	5-5.5	limonite
	5-5.5	lizardite, serpentine group
	5-5.5	magnetite
	5-5.5	talc
	5-6	anthophyllite
6-7	6	hemimorphite
	6-7	chloritoid
	6-7	variscite
7	7	alunite
	7	andalusite
	7	ankerite
	7	aragonite
	7	aurichalcite
	7	calcite
	7	cassiterite
	7	chalcedony
	7	chromite
	7	chrysoberyl
	7	chrysocolla

Fusibility[1]		Mineral[2]
	7	corundum
	7	diamond
	7	dioptase
	7	dolomite
	7	fayalite olivine
	7	forsterite olivine
	~7	glauconite, mica group
	7	graphite
	7	hematite
	7	illite, clay group
	7	ilmenite
	7	iron, meteoritic
	7	kaolinite, clay group
	7	kyanite
	7	leucite
	7	magnesite
	7	opal
	7	platinum, native
	7	pyrolusite
	7	quartz (normal)
	7	romanèchite
	7	rutile
	7	sillimanite
	7	montmorillonite, clay group
	7	smithsonite
	7	sphalerite
	7	spinel
	7	staurolite
	7	topaz
	7	turquoise
	7	uvarovite, garnet group
	7	zircon
not determined	ND	clinoatacamite
	ND	cuprorivaite
	ND	hollandite
	ND	huntite
	ND	palygorskite
	ND	pararealgar
	ND	paratacamite

Part II
Rocks

Chapter 4: Overview of Rocks

4.1 The Rock Cycle

When many grains of one or more minerals combine to form a solid aggregate (Figure 1.1), such a composite material is called a 'rock' and is given a unique name based on its origin, mineralogy, and grain size (a.k.a. texture), and also, sometimes, on its internal structure and grain shape. There are three classes of rock, each representing a fundamentally different origin: [1] igneous – forms through the cooling and solidification of magma or lava (i.e. molten or melted rock) or the accumulation of airborne debris during explosive volcanic eruptions; [2] metamorphic – forms when pre-existing rocks of any type are altered by high temperatures, high pressures, and/or chemically-active fluids; and [3] sedimentary – forms from the lithification (hardening) of loose sediments produced either by weathering and erosion of pre-existing rocks or by biological processes. All these rocks can be brought to the surface from great depths in the Earth's crust by faulting or uplift caused by the movement of tectonic plates. Once at the surface, the rocks are broken down into sediment, by weathering and erosion, and may eventually become soil. The interrelationships among the three rock types and the geological processes that affect them are embodied within what has been called the 'rock cycle' (Figure 4.1).

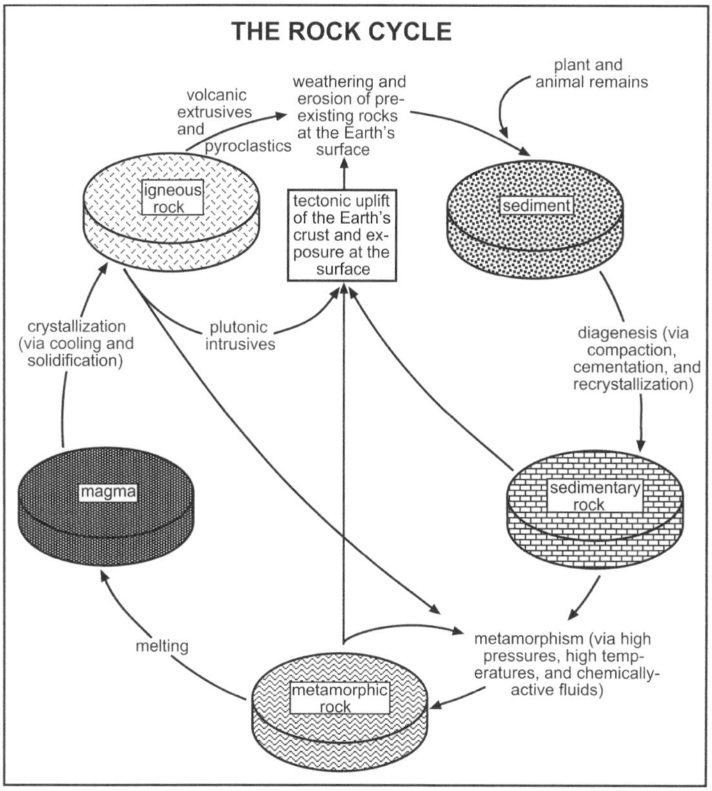

Figure 4.1: The rock cycle (from Harrell 2024: fig. 2.1).

4.2 Igneous Rocks

Most igneous rocks form through the crystallization of magma as it cools (i.e. the elements combine to form intergrown minerals). If crystallization occurs below the Earth's surface, the resulting rocks are said to be 'plutonic' or 'intrusive', and their slow cooling rate gives them larger grains. The term 'intrusive' derives from the fact that magma always moves upwards through the crust due to its lower density relative to the surrounding solid rocks and, consequently, it forces its way into the overlying, pre-existing (country) rocks. Intrusive igneous rock bodies have a variety of names that reflect their size and geometry. Sometimes magma works its way to the Earth's surface, where it exits through a volcanic cone or fissure, and such events are called 'volcanic eruptions'. When extruded onto the surface during quiet (effusive) eruptions, magma is referred to as 'lava'. Magma, along with pieces of solid rock, can also be blasted out of volcanoes and fissures during explosive eruptions and such airborne materials are known as 'pyroclasts'. Deposits form when these either fall to the surface downwind of an eruption (a 'pyroclastic fall'), or mix with expelled gases and flow down the flanks of a volcano as an incandescent, high-density cloud (a 'pyroclastic flow'). The 'volcanic' igneous rocks are, thus, those derived from pyroclastic falls and flows as well as lava flows, with the latter also referred to as 'extrusive' volcanic rocks (Figure 4.2). The rapid cooling rate of the extrusive rocks gives them grains that are much smaller than those in their plutonic cousins. Small, shallow intrusive bodies (e.g. dikes and volcanic pipes) associated with volcanic eruptions are commonly also called volcanic rocks, even though they are technically plutonic. These, however, are more appropriately referred to as 'subvolcanic' rocks. The true plutonic rocks are considered to be the larger, deeper intrusive masses or 'plutons' (batholiths, dikes, laccoliths, sills, and stocks in Figure 4.2), but also include the thin veins emanating from the plutons. Pyroclastic rocks, it should be noted, do not entirely conform to the common definition of igneous rocks: i.e. crystallization from magma. The individual grains (pyroclasts) are certainly formed from crystallized magma but the pyroclastic rock itself is an aggregate of grains that has been hardened into a solid mass (lithified) through compaction and cementation, just as in sedimentary rocks. Consequently, 'pyroclastic volcanic rocks' are sometimes also referred to as 'volcanoclastic sedimentary rocks'.

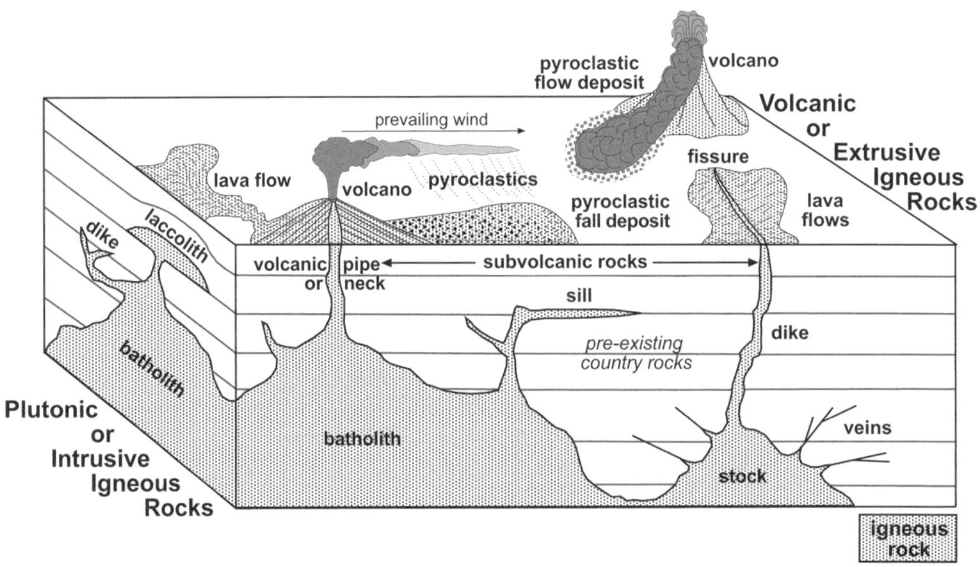

Figure 4.2: Spectrum of igneous rock occurrences (from Harrell 2024: fig. 2.2).

4.3 Sedimentary Rocks

4.3.1 Weathering, Sediments, and Depositional Environments

When rocks in the Earth's crust are exposed at the surface they begin to disintegrate. The disintegration process is referred to as 'weathering', and the rock and mineral fragments produced are called 'sediment'. There are two categories of weathering: 'mechanical' and 'chemical'. The former encompasses those physical processes that break a rock or mineral into fragments. It includes freezing and thawing of water in pores and fractures (frost wedging), growth of plant roots into cracks, expansive crystallization of halite (rock salt) in pores, burrowing activities of animals, thermal expansion and contraction during the daily heating cycle, and abrasion of fragments during transport. Chemical weathering, as the name implies, disintegrates rocks by chemical rather than physical means. The processes include: dissolution due to slightly acidic water or water that is under-saturated with respect to a mineral; oxidation in the presence of water due to a reaction between iron or other metallic elements in minerals with oxygen; and hydrolysis due to a decomposition reaction between silicate minerals and water. The feldspars and ferromagnesian minerals (i.e. those with iron and/or magnesian, such as augite, biotite, hornblende, and olivine) are especially susceptible to these processes. The ultimate stable by-products of chemical weathering are mostly quartz, clay minerals, and iron oxides.

Some of the sediment produced by weathering on the continental surfaces is carried to the oceans, but not all the sediment accumulating on the ocean floors is of this origin. Great quantities of sediment are derived from the remains of marine plants and invertebrate animals. Other marine sediments are the result of chemical reactions occurring as seawater becomes supersaturated with respect to calcium carbonate (e.g. ooliths), or when seawater evaporates, as it sometimes does when seas become geographically isolated from the world ocean (e.g. deposits of halite, anhydrite, and gypsum). The vast majority of sediment grains are derived from continents and are either mineral fragments (pieces of a single mineral) or rock fragments (pieces of an aggregate of two or more minerals). Both are referred to as 'siliciclastic' sediments as the grains consist mainly of silicate minerals.

Sediment is picked up (eroded) by rivers, winds, or glaciers on the continents, and by currents or waves in the oceans, and then transported many – sometimes hundreds of – kilometers before being permanently deposited in any of a wide array of sedimentary environments (Figure 4.3). Such sediment grains are said to be 'detrital' as they represent erosional detritus.

4.3.2 Diagenesis

Sediments become sedimentary rocks through the processes of compaction, cementation, and recrystallization, and these are collectively referred to as 'diagenesis' (Figure 4.4). Their effect is to harden or 'lithify' a loose sediment. Compaction results from the weight of overlying sediments which, for a given sediment layer, increases over time as deposition at the surface continues and the layer becomes progressively more deeply buried. The pore spaces between sediment grains are often filled with water, which usually has a number of dissolved ions in it. Under certain conditions, these ions can precipitate (i.e. come out of solution) in the pores, depositing minerals that act as a cement or natural glue that holds the sediment grains

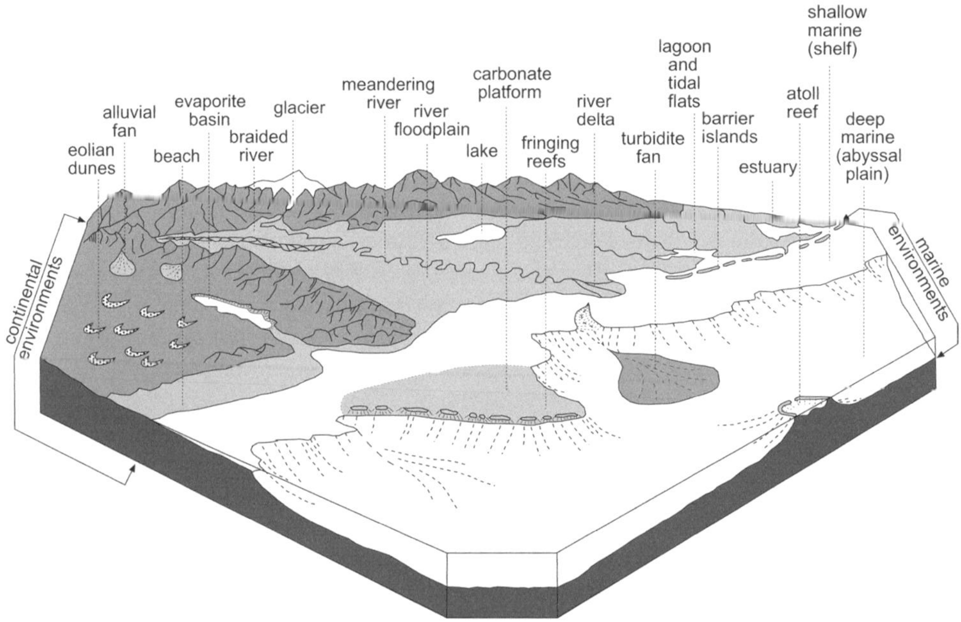

Figure 4.3: Spectrum of sedimentary depositional environments (from Harrell 2024: fig. 2.6).

together. Several minerals can form cements in sedimentary rocks and the most common of these are, in order of decreasing abundance, calcite, quartz, clay minerals, goethite/limonite, and hematite. Other sedimentary rocks are the result of secondary mineralization and recrystallization within pre-existing sedimentary rocks. Chert and dolostone replacing limestone, and (rock) gypsum replacing (rock) anhydrite are common examples of this process.

4.4 Metamorphic Rocks

Any pre-existing igneous, sedimentary, or metamorphic rock – collectively referred to as 'parent rocks' or 'protoliths' – will undergo significant changes in composition and/or texture when exposed to high temperatures, high pressures, or chemically-active fluids. This transformation process is called 'metamorphism' and normally takes place deep below the Earth's surface. There the confining (lithostatic) pressure steadily increases with depth because of the increasing weight of the overlying rocks. High pressures can also be applied laterally and at shallower depths where tectonic plates are converging or rocks are sheared along faults. Temperature steadily increases downward in the Earth, rising up to about 6000°C in the core. The heat escapes outwards, eventually reaching the Earth's crust where it is responsible for igneous activity and, at temperatures below the melting point of rocks, metamorphism. Such high metamorphosing temperatures can occur over broad regions of the crust. High temperatures can additionally occur more locally near the contact with igneous intrusions. As these magma bodies cool, they shed enormous amounts of heat into the surrounding country rocks. They also commonly produce hydrothermal fluids (both water-rich magmas and hot waters enriched in dissolved ions) that permeate the surrounding rocks and cause other changes. These water-based fluids can be derived from the magma itself, from heated ground water within the surrounding rocks, or from water expelled as a result of dehydration reactions during metamorphism.

From the above processes, three broad categories of metamorphism can be recognized: 'regional', 'contact', and 'hydrothermal' (Figure 4.4). Regionally metamorphosed rocks are produced by both high temperatures and pressures and, depending on the intensity of these processes, the changes can be characterized as low-, medium-, or high-grade metamorphism. At the shallowest depths, low-grade regional metamorphism passes upward into the realm of sedimentary diagenesis. Contact-metamorphosed rocks show the effects of primarily high temperatures and secondarily of chemically active fluids adjacent to a magma body. The affected zone varies in thickness from less than one meter beside small dikes to a few kilometers around stocks and batholiths. Hydrothermally metamorphosed rocks can occur anywhere in the subsurface where fluids can reach, but especially along faults and fractures, which act as conduits for the fluids. Two other, minor categories of metamorphism are recognized, 'dynamic' and 'shock', where the former is produced by shear stress directed along faults and the latter is caused by meteorite impacts. High pressure during mainly medium- and high-grade regional metamorphism is responsible for the development of 'foliation', a distinctive type of structure found only in metamorphic rocks. Non-foliated or, at most, weakly foliated rocks are produced by the other types of metamorphism as well as some low-grade regional metamorphism.

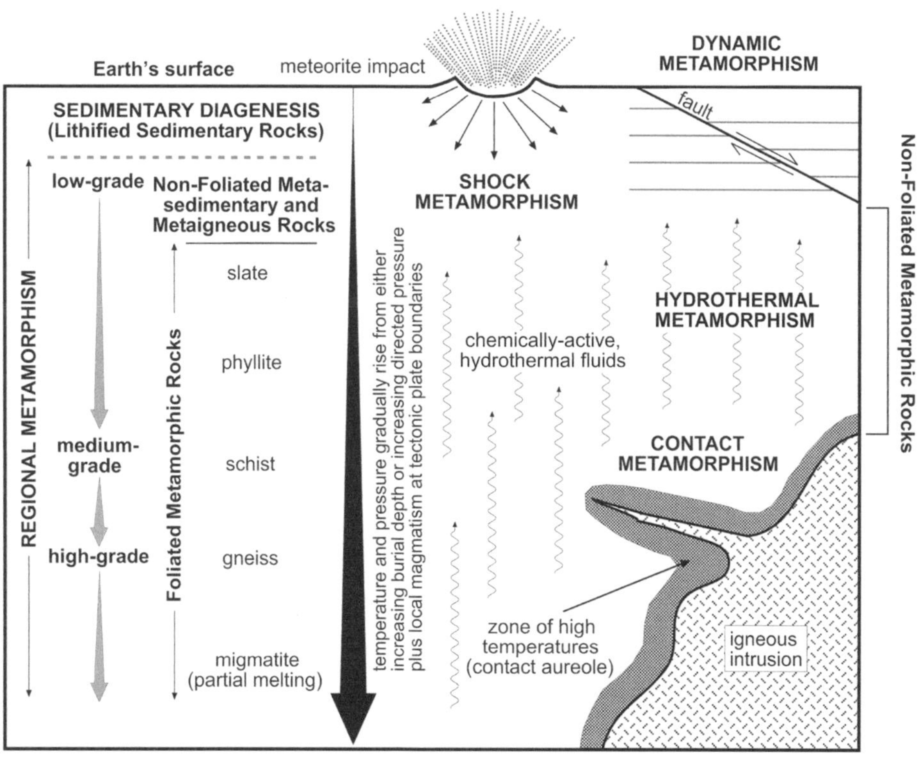

Figure 4.4: Spectrum of metamorphic rock occurrences (from Harrell 2024: fig. 2.7).

4.5 Archaeologically Important Rocks

Whereas only a tiny fraction (a little over 2%) of the known mineral varieties was employed in antiquity for various applications, essentially every kind of rock was put to use at one time or place in the past. In volumetric terms, the ordinary building stones had the greatest use, but in terms of the number of discrete objects recovered, those fashioned from utilitarian stones are often the most plentiful at archaeological sites. More memorable and celebrated, however, are the ornamental stones with which ancient cultures decorated their buildings and tombs, and carved their most important objects. Although the gemstones employed for jewelry in antiquity are mostly minerals (see section 2.2), some are also rocks. Summarized below are the principal uses of rocks, the properties that make them suitable for a particular application, and the rock types most commonly encountered.

Building Stones

Application 1: solid rocks carved into ashlar blocks and slabs for walls, pavements, doorways (i.e. thresholds, jambs, and lintels), window frames, columns and pillars (including capitals and bases), architraves and pediments, roofing slabs, and other architectural and structural elements. Some structures, especially walls and pavements, were also built using stacked cobbles, boulders, or other unworked pieces of rock either with or without mortar.

> *Attributes and rock types:* any readily available, relatively easy to quarry and carve, and structurally sound rock. The variety employed varied widely depending on local availability and technological level. The most common varieties include limestone, sandstone, rock gypsum, marble, and volcanic tuff. Rocks of all kinds, however, were employed in the stacked-stone constructions. Muddy sediment rich in clay minerals was shaped and dried into mud bricks, and used as an alternative to stone ashlars.

Application 2: plaster applied over stone surfaces as a painting substrate and mortar applied between stone blocks as a binder.

> *Attributes and rock types:* heating at a high temperature (i.e. calcination) reduces limestone and marble to quick lime, and rock gypsum to calcium sulfate hemihydrate (a.k.a. burnt gypsum or plaster-of-Paris). The addition of water to these powdered materials produces, respectively, lime plaster/mortar and gypsum plaster/mortar. Muddy sediment rich in clay minerals was also used for plaster and mortar.

Ornamental Stones

Miscellaneous applications: statuary and other sculptures, funerary and other vessels, basins, stelae, megalithic standing stones, sarcophagi, small shrines, alters and offering tables, as well as decorative building elements such as wall facings or veneer, patterned pavements, columns and capitals, and door and window frames.

> *Attributes and rock types:* any rock with an attractive color or pattern which is at least moderately hard and structurally sound. Typical examples from antiquity are travertine and other crystalline limestones, basalt, silicified sandstone, quartzite, marble, granite,

and other crystalline igneous and metamorphic rocks. Sometimes the primary appeal of a rock for an ornamental application was its ease of carving, rather than durability. Consequently, rocks more commonly employed as building stones were sometimes used in this way. Another soft stone occasionally used for ornamental purposes was steatite.

Utilitarian Stones

Application 1: sharp-edged cutting tools and weapons, including knives and daggers, projectile and spear points, axes, sickle teeth, awls, drill bits, burins, adzes, and scrapers.

Attributes and rock types: any hard, brittle, fine-grained, homogenous rock that can hold a sharp edge produced by either knapping or grinding. The best-known materials are chert/flint, jade, and obsidian, but other rock varieties commonly used are vein quartz, petrified wood, metamorphic quartzite and hornfels, volcanic igneous rocks (e.g. basalt, dacite, rhyolite, and welded tuff), and silicified sedimentary rocks (e.g. quartz-cemented or silicified sandstone, and quartz-impregnated or siliceous limestone, dolostone, and mudrock).

Application 2: blunt-edged percussive tools and weapons, including hammerstones or pounders, maces and mauls, hoes, gouges, and plowshares.

Attributes and rock types: any hard stone with good impact resistance. This includes nearly all igneous and metamorphic rocks – when they are unweathered and unfractured – as well as some harder sedimentary rocks (e.g. recrystallized limestone and dolostone, and silicified sandstone).

Application 3: milling and grinding stones for cereals and other plant products, mineral pigments, and metallic ore rocks.

Attributes and rocks types: any hard, fracture-resistant igneous, metamorphic, or sedimentary rock. Rocks with a mixture of hard and soft areas on their grinding surfaces (i.e. hard grains alternating with soft matrix or voids) were particularly favored for processing plant materials as the harder grains stand out, enhancing their abrasiveness (e.g. vesicular basalt, porous silicified sandstone, and calcite-cemented quartz sandstone).

Application 4: cooking vessels.

Attributes and rock types: an easily carved, fire-resistant rock, principally steatite.

Other applications:

(i) abrasives – grits for grinding, drilling, and polishing hard rocks (mainly quartz sand and the emery form of corundum), rubbing stones for smoothing and polishing hard rocks (especially silicified sandstone) or rough skin (pumice), and hone or whetstones for sharpening metal tools and weapons (especially chert/flint and silicified sandstone)
(ii) ship anchors and ballast (any rock type for ballast but easily carved, water-resistant ones for anchors)

(iii) weights and measures (any moderately hard, easily carved rock that is resistant to chemical reactions and wear that can change its weight)
(iv) oil lamps, pipe bowls, and molds for metal castings (any easily carved, thermally resistant rock, especially steatite, sandstone, and sepiolite)
(v) fuel (coal and lignite)
(vi) lithographic printing (limestone)
(vii) paint pigments (e.g. white from powdered limestone and diatomite, blue from lapis lazuli, and a variety of colors from other mineral aggregates, as noted in Tables 3.4-3.5)
(viii) ceramics, including pottery, crucibles for metal smelting and refining, fired bricks, drainage and smoking pipes, floor and roofing tiles, loom weights and spindle whorls, stamped bullae, and figurines and some larger sculptures (manufactured from sedimentary mudrock or sediment rich in clay minerals)
(ix) tablets for cuneiform writing (damp mud that is rich in clay minerals)
(x) stone bearings ('tournettes') for potters' wheels (any hard stone)
(xi) strike stones for starting fires (either pyrite or marcasite, which produce a cascade of sparks when struck with a harder stone such as chert/flint or vein quartz)

Gemstones

Miscellaneous applications: jewelry (beads and pendants, and inlays in metal cloisonné frameworks) and small decorative objects (amulets, seals, small vessels, and statuettes).

> *Attributes and rock types:* any relatively hard, attractive rock or mineral (or other natural substance) can be considered a 'gemstone'. Most gemstones are minerals (see section 2.2), but some rocks were also employed as gemstones, including lapis lazuli, obsidian, serpentinite, hardened steatite, and travertine. Other gemstones are also technically rocks as they occur as granular aggregates, but are usually treated as minerals (e.g. agate and other chalcedonies, and jade). Many of the rocks used for ornamental applications were also occasionally employed for jewelry, especially beads and pendants.

There is one aspect of the study of archaeological stones where the archaeologists are out ahead of the geologists. This is the subdiscipline of 'lithics', which involves the analysis of knapped (chipped) or ground stone tools and their debitage (see Odell 2004 and Andrefsky 2005 for good overviews). The identification of stones is usually of secondary importance to the understanding of the forms and functions of the tools and their manufacturing techniques, but there are some exceptions (e.g. Clough and Woolley 1985; Luedtke 1992).

4.6 Rock Identification

The identification process is generally more difficult for rocks than for minerals. This is both because rocks usually consist of more than one type of mineral (commonly several), and because it is necessary to know the relative amounts of the minerals in order to make use of the various rock-classification schemes. Further adding to the difficulty, a given 'classified' rock variety can be variable in its mineralogy, texture, and appearance. Nevertheless, with careful observation, it is possible to identify rocks by macroscopic means or, at least, greatly narrow the possibilities for their identity. There is essentially no disagreement on how to identify mineral species, although different names are sometimes applied, but for rocks there are competing classifications where each recognizes a specific rock type – e.g. 'granite' – as something slightly different.

There is only one wholly reliable methodology available to petrologists for identifying rocks: 'thin-section petrography'. This involves point-counting mineral grains in thin sections using a petrographic microscope equipped with a mechanical stage placed between rotatable polarizing filters. A variation on this technique ('optical mineralogy') was noted in section 2.3 in connection with the identification of minerals. For rocks, however, the addition of a mechanical stage allows the petrologist to identify the mineral at each of several hundred points across the thin section and so determine their relative amounts. Thin-section petrography also permits the recognition of textures, internal structures, and mineral interrelationships that are additionally needed to identify a rock. For the volcanic igneous rocks, which are difficult to study in thin section because of their especially fine grain size or glassy composition, geochemical analyses are also required. In comparison, much less information is available from a macroscopic examination.

A precise, detailed identification of rocks by macroscopic means is seldom possible, but approximate identifications are usually achievable. Chapter 5 describes how this can be done. The classification schemes required for this have been simplified and, for competing schemes, harmonized to include properties that can be observed, measured, or estimated with a hand-lens and some simple tests. This will suffice to identify about 99% of the rock varieties you might potentially encounter. For the other 1% not included in the classification tables, you should still be able to say something about the gross mineralogy and texture, and, thereby, recognize the rock class and perhaps also the broad category of rock type within a class.

4.7 Key for Determining Rock Class

Before you can apply any of the rock classifications, you need to determine, at least tentatively, which rock class your stone specimen belongs to: i.e. is it igneous, metamorphic, or sedimentary? Herein lies a problem. When geology students are learning about rocks and minerals, they study minerals first and then, after gaining some expertise in mineralogy, they attempt to identify rocks. Archaeologists and other non-geologists will not normally have this mineralogical background, and this makes identifying the rock class especially challenging. The coarse-grained rocks (i.e. those with grains visible to the unaided eye or with a hand-lens) are naturally the easiest to work with. Major difficulties are encountered, however, with the fine-grained rocks where things are more difficult to see. Nevertheless, using the key below, you can almost always reduce the possibilities to at least two rock classes and often to only one. If you are undecided between two classes, then go through the identification process for both as described in Chapter 5. One of the outcomes will probably fit your specimen better than the other. Before using the key below, you need to understand some of the terminology it employs.

Terminology Used in the Rock Class Key

Macroscopically discernable grains: individual grains can be seen with the unaided eye (or with reading glasses). This corresponds to the 'phaneritic' texture in igneous and metamorphic rocks, and 'sand' and coarser grain sizes in sedimentary rocks. If the grains are mostly or entirely non-discernable, this corresponds to the 'aphanitic' texture in igneous and metamorphic rocks, and 'silt' and finer grain sizes in sedimentary rocks. Some geologists refer to these, respectively, as 'coarse-' and 'fine-' grained textures.

Crystalline texture: this consists of a mosaic of interlocking (i.e. intergrown) mineral grains of various sizes, shapes, and colors (Figure 4.5a; see also Figures 5.1-5.2 and 5.12-5.14). It occurs in all rock classes but is most common in igneous and metamorphic rocks. As described in section 1.2, you may have to wet the rock's surface with water, spittle, or mineral oil to clearly see the intergranular relationships.

Clastic (a.k.a. fragmental) texture: this consists of originally loose grains (clasts) that are now cemented together (Figure 4.5b; see also Figures 5.5 and 5.7-5.8). It occurs primarily in sedimentary rocks that consist of 'detrital' grains (i.e. those transported and deposited by water, wind, or ice), but is also found in igneous rocks composed of 'pyroclasts' (i.e. glass, and crystal and rock fragments blasted out of a volcano). Incompletely metamorphosed sedimentary rocks, the so-called 'metasediments', are low-grade metamorphic rocks that may have a relict clastic texture.

Banded structure: the rock has a crystalline texture with the grains segregated into alternating bands of light- and dark-colored minerals that can be straight, wavy, or crinkly (Figure 4.5c; see also Figure 5.11). This occurs in the so-called 'gneissose' metamorphic rocks where it is recognized as one type of 'foliation', but on rare occasions something similar occurs in igneous rocks, where layers of different minerals form as these sequentially precipitate and settle-out in a magma body.

Directional structure: the rock has a crystalline texture with the elongated and/or platy grains exhibiting a parallel orientation (Figure 4.5d; see also Figures 5.9-5.10). This is another type of foliation found in many metamorphic rocks, but on rare occasions something similar occurs in igneous rocks, a 'preferred orientation' in the larger grains produced by flowing magma.

Scratch hardness: this is the Mohs hardness (see section 3.2.3.1). Although this measure of strength really only applies to single minerals, it is a useful way to distinguish between hard and soft rocks when the test is done using the tip of a knife blade or other steel tool with a hardness of about 5.5 (i.e. ordinary steel). If a majority of the rock consists of mineral grains with a Mohs hardness over 5.5 and these are tightly bound, a metallic streak will be left from the steel but little or no groove will be cut (i.e. it is a 'hard rock'). If, however, the rock consists of either softer minerals or harder grains that are weakly aggregated, little or no streak will be left but a groove will be cut (i.e. it is a 'soft rock'). Knowing why a rock is soft can be helpful for its identification. If you suspect that the grains are merely being pushed aside by the tool rather than scratched, try using the rock to scratch the tool. If it contains minerals harder than steel, even if the grains are loosely held, the rock will scratch the metal. A broad knife blade works best for this test.

Glassy, granular, and porcelaneous surfaces: these are notable properties of fine-grained rocks and are determined by rubbing your finger along the surface and also by observing the surface luster. One group of igneous rocks is composed largely or entirely of glass and so have surfaces that are shiny (highly vitreous) and very smooth, although pervasively rippled due to conchoidal fracturing (Figure 4.5e). The broken edges on these rocks are exceedingly sharp. Some sedimentary rocks will have a dull to somewhat shiny porcelaneous (i.e. resembling unglazed porcelain) surface with a rougher feel, waxy luster, and only moderately sharp edges. Other igneous rocks, along with some metamorphic rocks, will have non-reflective surfaces that feel rough due to the rock's fine granularity and will have mostly dull edges.

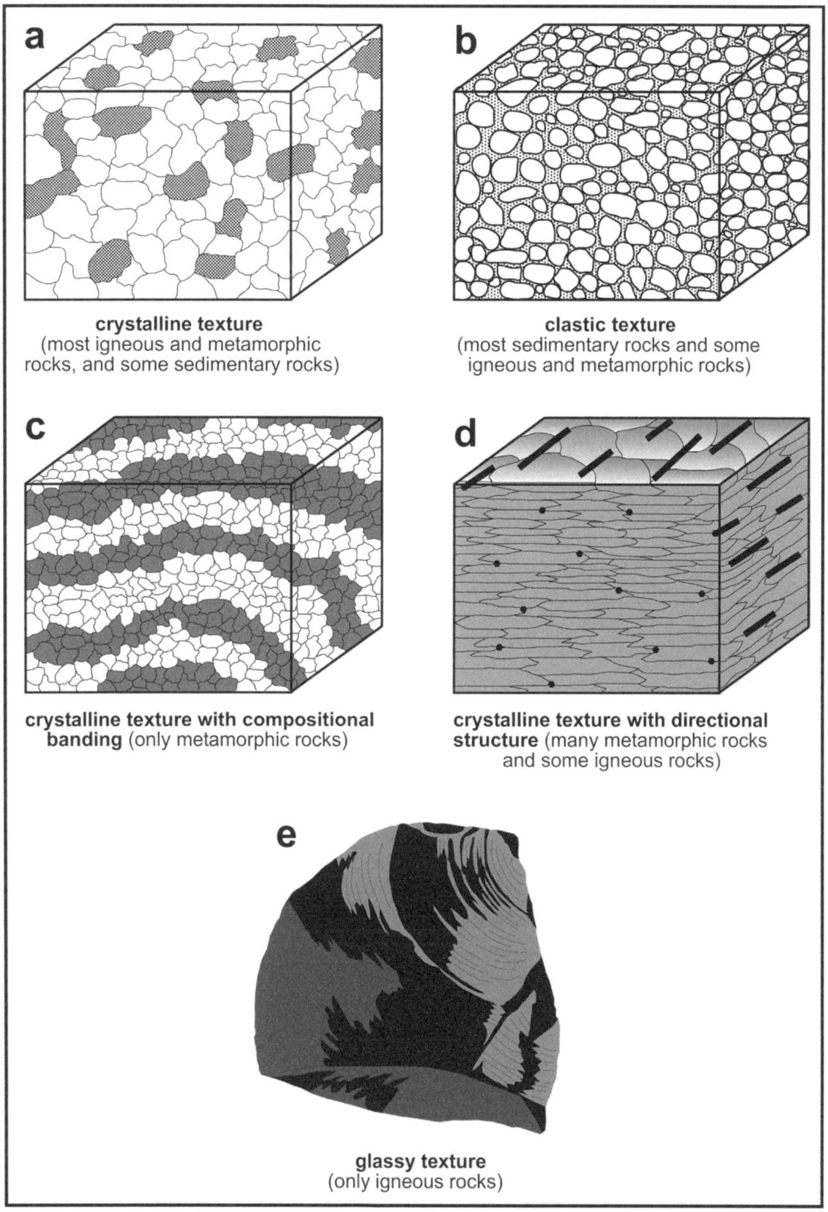

Figure 4.5: Basic textures for igneous, metamorphic, and sedimentary rocks.

ROCK CLASS KEY

It is not an easy matter to condense into a series of steps the thought processes that might go through a geologist's mind when seeing a rock specimen for the first time. What follows is an attempt at such a distillation.

1. Does the rock have grains that are mostly macroscopically discernable?
   If yes, go to 2.
   If no, go to 14.
2. What kind of texture does the rock have?
   If crystalline, go to 3.
   If clastic, go to 11.
   If organic framework, it is SEDIMENTARY.
3. Can the rock (i.e. its major constituents) be scratched by ordinary steel?
   If yes, go to 6.
   If no, go to 4.
4. Does the rock have a directional structure?
   If yes, it is probably METAMORPHIC, but possibly IGNEOUS.
   If no, go to 5.
5. Does the rock have a banded structure?
   If yes, it is probably METAMORPHIC, but possibly IGNEOUS.
   If no, it is probably either IGNEOUS or METAMORPHIC, but it is more likely the former if the rock has an overall pinkish or reddish color.
6. Can the rock (i.e. its major constituents) be scratched by a fingernail?
   If yes, go to 7.
   If no, go to 8.
7. Is the rock light or dark colored?
   If light, it is probably SEDIMENTARY, but possibly IGNEOUS or METAMORPHIC.
   If dark, it is METAMORPHIC.
8. Does the rock have a directional structure?
   If yes, it is METAMORPHIC.
   If no, go to 9.
9. Does the rock vigorously effervesce in dilute HCl acid when either whole or powdered?
   If yes, it is probably either METAMORPHIC or SEDIMENTARY, but possibly IGNEOUS.
   If no, go to 10.
10. Does the rock have a salty, bitter, or astringent taste?
    If yes, it is SEDIMENTARY.
    If no, go to 11.
11. Does the rock contain fossils?
    If yes, it is SEDIMENTARY.
    If no, go to 12.
12. Can the rock (i.e. its major constituents) be scratched by ordinary steel?
    If yes, it can be either IGNEOUS, METAMORPHIC, or SEDIMENTARY. It is sedimentary if the rock has spherical or ovoid grains, or a dark-brown, brownish-black, or black color.
    If no, go to 13.
13. Does the rock break around or through the grains?
    If around, it is SEDIMENTARY.

If through, it is METAMORPHIC.

14. Can the rock (i.e. its major constituents) be scratched by ordinary steel?
    If yes, go to 19.
    If no, go to 15.
15. Does the rock have abundant open cavities, with or without secondary mineral deposits in them?
    If yes, it is probably IGNEOUS, but possibly SEDIMENTARY.
    If no, go to 16.
16. Does the rock consist of large, macroscopically discernable grains surrounded by a matrix of smaller, non-discernable grains?
    If yes, it is probably IGNEOUS.
    If no, go to 17.
17. Does the rock have a smooth or rough/grainy surface?
    If smooth, go to 16.
    If rough/grainy, it can be either IGNEOUS or METAMORPHIC, but it is more likely the former if the rock has an overall pinkish or reddish color.
18. Does the rock's surface have a glassy or porcelaneous luster?
    If glassy, it is IGNEOUS.
    If porcelaneous, it is SEDIMENTARY.
19. Does the rock break into thin, platy fragments?
    If yes, go to 20.
    If no, go to 21.
20. When struck, does the rock emit a sharp ringing sound or a dull thud?
    If sharp, it is METAMORPHIC.
    If dull, it is SEDIMENTARY.
21. Does the rock have a greasy feel or a blue color?
    If yes, it is METAMORPHIC.
    If no, it is probably SEDIMENTARY, but possibly METAMORPHIC.

Chapter 5:
Rock Properties and Identification

5.1 Equipment and Supplies Needed

1. Hand-lens.
2. Knife blade or other piece of steel to test for scratch hardness.
3. 5% or 10% HCl acid to test for carbonate minerals.
4. Strong (neodymium) magnet to test for magnetic minerals.
5. Equipment to measure specific gravity.
6. A ruler or other scale with millimeter divisions.

5.2 Igneous Rocks

5.2.1 Description

Most igneous rocks have a crystalline texture (i.e. they consist of interlocking mineral grains that grew together during crystallization from magma; Figure 4a). If the majority of individual grains are large enough to be seen with the unaided eye (i.e. larger than about ~0.5 mm), the texture is said to be 'phaneritic' and the rock is a 'phanerite' (Figure 5.1). This assumes, of course, that you have good eyes or your vision is corrected by glasses. If the grains are mainly too small to be clearly seen, then the texture is described 'aphanitic' and the rock is an 'aphanite'. Plutonic rocks are always phaneritic, whereas volcanic rocks are aphanitic or 'glassy' (Figure 4e) when extrusive, aphanitic to finely phaneritic when subvolcanic intrusives, or have a non-crystalline 'clastic' texture (Figure 4b) when pyroclastic. Volcanic aphanites are not always wholly crystalline. They may contain significant amounts of glass that is not macroscopically discernable. Some intrusive masses, especially dikes, have exceptionally large grains due to the abnormally high water content of the magma, and these are said to exhibit a 'pegmatitic' (i.e. very coarsely phaneritic) texture with the rock commonly referred to as a 'pegmatite'. The mineral grains in such rocks are at least a few centimeters across and can even be up to an astounding several meters in maximum dimension. Conversely, water-poor plutonic dikes usually have a finely phaneritic texture (i.e. grains ~1-2 mm) with the rock referred to as 'aplite' or 'dolerite' (or 'diabase') depending on the composition (see Tables 5.3 and 5.5, respectively). Although 'aplite' has a compositional connotation, it is non-specific; consequently, it as well as the strictly textural term 'pegmatite' will have a compound name based on a combination of their mineralogy and texture: e.g. pegmatitic diorite and granite aplite. Volcanic and subvolcanic rocks commonly, and plutonic rocks occasionally, will have grains of two widely different sizes and this results from an initial period of slower cooling (forming the larger grains, the 'phenocrysts') followed by more rapid cooling (resulting in smaller grains, the 'matrix' or 'groundmass'). This is known as a 'porphyritic' texture. The phenocrysts in a given rock are usually just one variety of mineral but occasionally two types of phenocrysts will occur together. Phenocrysts usually exhibit the mineral's crystal form, although rarely perfectly, and so are relatively easy to identify. On rare occasions, elongated or tabular phenocrysts can exhibit a preferred orientation due to flow effects within a viscous magma. Both the phenocrysts and matrix are phaneritic in plutonic rocks, but volcanic rocks have phaneritic phenocrysts set in an aphanitic matrix (Figure 5.2). Porphyritic rocks are given a compound name that indicates both the texture and composition: e.g. andesite porphyry or, interchangeably, porphyritic andesite.

Phenocrysts are not the only grains that can occur as well-formed ('euhedral') crystals in igneous rocks. The earlier formed (higher temperature) minerals in non-porphyritic phanerites also commonly have a good crystal shape. These, like the phenocrysts, had the freedom to grow relatively unhindered in a magma whereas the minerals that formed later at lower temperatures could only grow into an increasingly more restricted space between the earlier-formed crystals. Which minerals crystallize early depends on the composition of the magma. For example, it is the potassic and plagioclase feldspars in silica-rich magmas, and augite (or other pyroxenes), plagioclase feldspar, and olivine in silica-poor magmas (see section 5.2.2 for magma types). Hornblende (and other amphiboles) and biotite mica crystallize at intermediate temperatures and sometimes are able form partially developed ('subhedral') crystals, but quartz, which is always the last mineral to crystallize from a cooling magma, occurs as irregular-shaped ('anhedral') grains. The only exception to this is in some felsic volcanic rocks where quartz occurs as well-formed phenocrysts.

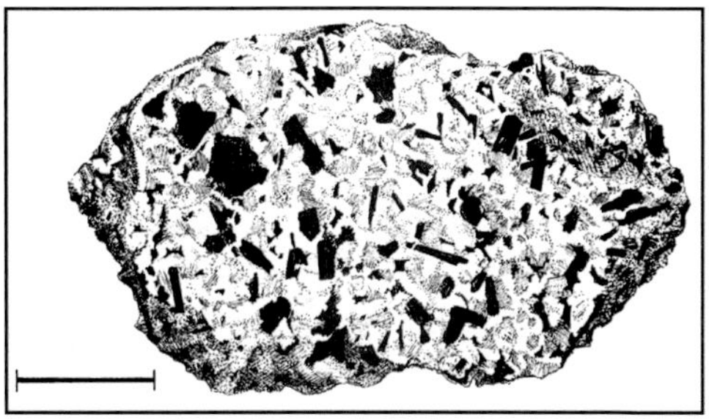

Figure 5.1: Example of a crystalline phanerite texture in an igneous rock (granodiorite). From Figure 10A in Raymond (1984) with permission of Waveland Press, Inc. Scale bar = 2 cm.

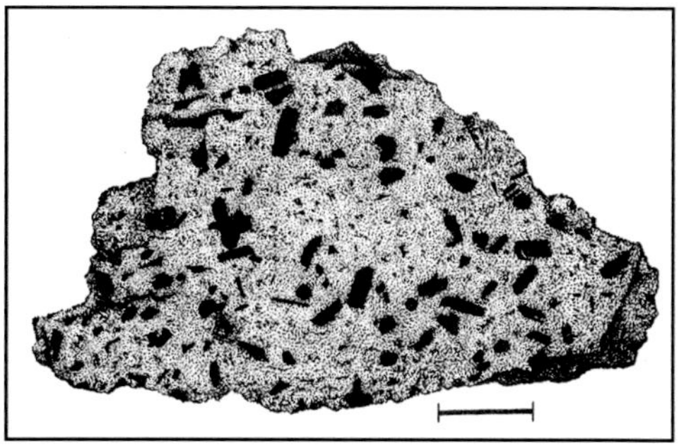

Figure 5.2: Example of a crystalline, porphyritic aphanite texture in an igneous rock (porphyritic andesite). From Figure 11A in Raymond (1984) with permission of Waveland Press, Inc. Scale bar = 2 cm.

The clastic (a.k.a. fragmental) textures of pyroclastic rocks are similar to those found in many sedimentary rocks (Figure 4.5b). The 'pyroclasts', which are rock and mineral fragments and pieces of glass, are not intergrown as in a crystalline texture, but rather are compacted and cemented together. The names applied to these rocks are based only on the predominant size of the pyroclasts. Lava flows, especially in their upper part, sometimes cool so rapidly that minerals do not have time to crystallize. The resulting rock is a kind of glass and so exhibits a glassy (amorphous) texture (Figure 4.5e). The last igneous texture of importance, 'vesicular', refers to those extrusive volcanic rocks that are full of holes (vesicles) that were formed by gas bubbles trapped in the rapidly cooling and solidifying lava (Figure 5.3). The vesicles are commonly the sites of secondary mineral deposits ('amygdules') with the rock then described as 'amygdaloidal'.

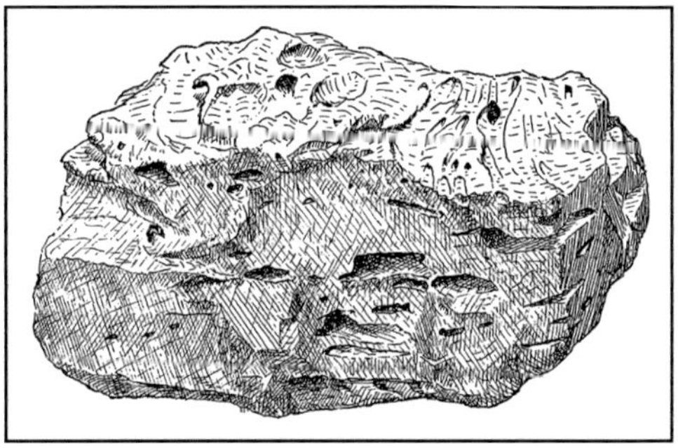

Figure 5.3: Example of a vesicular texture in basalt. From Figure 4-6 in Spock (1962) with permission of Harper and Row Publishers.

5.2.2 Classification

Many classifications of igneous rocks have been in use since the 19th century. These have produced competing definitions for some common rock names and have also introduced a multitude of new rock names unique to a particular classification or sub-field of geology. This situation has been a source of confusion and annoyance for geologists as well as archaeologists. Although common rock names like granite and basalt, for example, are defined in similar terms in all the classifications, their definitions in each scheme can differ in the details. One classification is more widely adopted internationally than all the others and this comes from the Subcommission on the Systematics of Igneous Rocks, organized under the auspices of the International Union of Geological Sciences or IUGS (Le Maitre 1989). A detailed discussion of the macroscopic application of this classification is provided by Dietrich and Skinner (1979: 111-169).

The IUGS classification of phaneritic (plutonic) rocks is done on the basis of the relative amounts of the most common minerals or mineral groups: i.e. [1] quartz; [2] the alkali feldspar group (including the potassic feldspars – mainly orthoclase and microcline but occasionally also anorthoclase or sanidine – plus albite plagioclase); [3] the plagioclase feldspar group (the other five plagioclase varieties, oligoclase through anorthite); [4] the dark ferromagnesian minerals (most commonly augite, biotite, hornblende, magnetite, and olivine); and, in the silica-poor rocks, [5] the feldspathoid minerals (most commonly nepheline) that take the place of quartz and sometimes also feldspar. These minerals and their relative amounts are best determined by thin-section petrography (see section 4.6). The IUGS classification of aphanitic (volcanic) rocks, in contrast, requires both geochemical and petrographic analyses for the groundmass and phenocrysts, respectively. For those situations where only macroscopic observations are possible, the IUGS offers a 'preliminary' or 'field' version of its classification of igneous rocks (Le Maitre 1989: 22 and 31) and this is shown in Table 5.2. This divides the rock varieties into families with a similar composition and appearance. It is this simplified classification that is recommended for archaeologists. The minerals can be identified macroscopically using the tips in Table 5.1 (but see also Tables 3.4-3.6), and their relative amounts can be estimated with the aid of the visual comparator in Figure 5.4. To use the comparator, you match the apparent amount of a given mineral or mineral group in your specimen to one of the panels in the figure or interpolate between two panels (the percentages represent the total area of the rectangular shapes in a panel).

The most important portions of the detailed IUGS classifications of phanerites and aphanites, where individual rock varieties are defined, are presented in Tables 5.3-5.7. You should be able to easily determine which rock family an unknown igneous specimen belongs, but in some cases you can do better than this with the identification of a specific rock variety within a family. The simple classifications of glassy and pyroclastic igneous rocks in Table 5.8 are also part of the IUGS system. The relative global abundances of the rock varieties in Tables 5.3-5.8 are based largely on Streckeisen (1967: 236) and are indicated by the font used for their names as explained in a note below each table. The most frequently encountered rock varieties among those considered to be 'abundant' are granite, andesite, and basalt, and the most common family of phaneritic rocks is the granitoids.

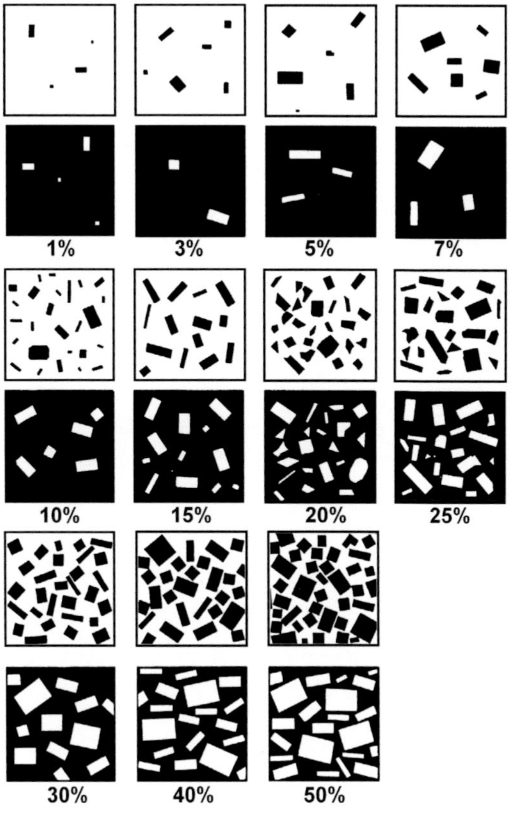

Figure 5.4: Visual comparator for estimating relative amounts of minerals in a rock. Adapted from Figures 13 and 14 in Raymond (1984) with permission of Waveland Press, Inc.

The crystalline phaneritic and aphanitic igneous rocks, and also the amorphous glassy ones, are conventionally placed into four groupings having both compositional and color connotations.

[1] 'Felsic' rocks contain abundant light-colored minerals (mainly quartz and feldspars). Felsic is a mnemonic adjective derived from feldspar and silica (i.e. SiO_2). These are roughly equivalent to the so-called 'acid' or 'acidic' rocks, with this name an indirect reference to their high silica content as silica reacts with water to form silicic acid (H_4SiO_4).

[2] 'Mafic' rocks contain abundant dark-colored, ferromagnesian minerals (mainly augite and other pyroxenes, biotite, hornblende and other amphiboles, magnetite, and olivine). Lighter colored plagioclase feldspar is usually common but quartz is rare or absent. Mafic is a mnemonic adjective derived from magnesium and ferric. These are roughly equivalent to the so-called 'basic' rocks, an indirect reference to their low silica content.

[3] 'Intermediate' rocks fall between the felsic and mafic ends of the compositional spectrum and often exhibit a mottled mix of light and dark minerals.

[4] 'Ultramafic' rocks consist almost entirely of dark-colored minerals (amphiboles, magnetite, olivine, pyroxenes, and/or phlogopite mica). These are also called 'ultrabasic' rocks.

These compositional adjectives are also commonly applied to the light- and dark-colored minerals themselves, as in Table 5.1. The overall color of a rock is a useful, but not infallible, indicator of mineralogy. It is helpful in identifying the phaneritic rocks, but is essential for the aphanitic ones where, apart from phenocrysts, the mineral grains are too small to be seen. The IUGS classification provides a Color Index (CI) as a measure of gross mineralogy for the felsic to mafic phaneritic rocks. It is a darkness scale that ranges from 0 (solid white) to 100 (solid black), and is roughly equivalent to the amount of dark-colored mafic minerals (as a percent of the whole rock) normally present in a specific rock variety. CI values are reported in Tables 5.3-5.6, but it must be kept in mind that the actual amount of mafic minerals in a given specimen will occasionally fall outside the usual range for a rock variety and so make it appear abnormally light or dark.

The IUGS does not provide Color Index values for aphanitic, glassy, or pyroclastic rocks. Nevertheless, the values given for phanerites still apply generally to the compositionally equivalent aphanites. More useful for identifying the latter, however, is the mineralogy of the phenocrysts as described in Table 5.9. While phenocrysts are very common, they are not always present and, consequently, color is sometimes the only means of identifying the aphanitic rocks. Another source of information that can assist in the identification of many igneous rocks is specific gravity (SG). The SG values reported in Tables 5.3-5.5 and 5.7-5.8 represent either the normal global range or average for each rock type and come from Clark (1966: Tables 4-1, 4-2. 4-3, and 9-2). Unfortunately, this parameter is available only for the more common rock varieties. Specific gravity can be used to reliably place an unknown igneous rock into one of the following three groups: [1] < 2.4 g/cc – glassy rocks; [2] 2.4-3.1 g/cc – felsic to mafic phaneritic and aphanitic rocks, and pyroclastic rocks (note that the SG value of the pyroclastics can drop below 2.4 g/cc when they are highly porous); and [3] > 3.1 g/cc – ultramafic rocks.

Small amounts of magnetite are nearly ubiquitous in crystalline igneous rocks and, hence, their detectable magnetism is not an aid to their identification. The glassy and pyroclastic rocks, on the other hand, almost always lack magnetite and so are non-magnetic. Among the igneous rocks, only the pyroclastic ones will have a scratch hardness below that of steel, although there is an exception. Sometimes during deposition from a pyroclastic flow, the temperature is so high that the grains are wholly or partially fused together, yielding a much harder rock known as a 'welded tuff'.

5.2.3 Identification Key

The following key, when used together with Tables 5.1-5.9, will allow you to identify the rock family, and sometimes a specific rock variety, for an unknown igneous specimen.

1. What is the rock's texture?
   If it is phaneritic and crystalline, go to 2.
   If it is aphanitic and crystalline, go to 8.
   If it is glassy, go to 10.
   If it is clastic, go to 11.
2. What is the quartz content?
   If quartz is common to abundant and conspicuous, go to 3.
   If quartz is rare or absent, go to 4.
3. What is the potassic feldspar content?
   If potassic feldspar is common and conspicuous and the rock has an overall pinkish, reddish, yellowish, or grayish coloration, it is either alkali-feldspar granite, granite, or granodiorite (see Table 5.3 to distinguish among these), or alkali-feldspar quartz syenite, quartz syenite, or quartz monzonite (see Table 5.4 to distinguish among these).
   If potassic feldspar is rare or absent and the rock has an overall grayish coloration, it is either granodiorite or tonalite (see Table 5.3 to distinguish between these) or quartz monzodiorite, quartz diorite, quartz monzogabbro, quartz gabbro, or quartz anorthosite (see Table 5.5 to distinguish among these).
4. What is the potassic feldspar content?
   If potassic feldspar is common and conspicuous, the rock is either alkali-feldspar syenite, syenite, monzonite (see Table 5.4 to distinguish among these), monzodiorite or monzogabbro (see Table 5.5 to distinguish between these), or their feldspathoid-bearing/rich equivalents (Tables 5.4-5.6).
   If potassic feldspar is rare or absent, go to 5.
5. What are the plagioclase and mafic mineral contents?
   If plagioclase constitutes all or nearly all of the rock, it is quartz anorthosite or anorthosite (see Table 5.5 to distinguish between these).
   If dark mafic minerals are less than 50% of the rock and light-colored plagioclase and/or feldspathoid minerals are abundant and conspicuous – giving the rock a light- and dark-speckled or mottled appearance, it is probably either diorite (Table 5.5) or its feldspathoid-bearing/rich equivalent (Tables 5.5-5.6).
   If dark mafic minerals are over 50% of the rock but less than 90%, it is almost certainly either gabbro (Table 5.5) or its feldspathoid-bearing/rich equivalent (Tables 5.5-5.6).
   If dark mafic minerals constitute nearly all of the rock (>90%), it is an ultramafic (Table 5.7). Go to 7.
   If the rock is light colored and not obviously one of the above, go to 6.

   > Note: the only compositional difference between dioritoids and gabbroids is the An content of the plagioclase: <50 for dioritoids and >50 for gabbroids. The different plagioclase minerals cannot be reliably identified by macroscopic means but there are common tendencies: plagioclase with An <50 is often white or light- to medium-gray whereas plagioclase with An >50 is often dark- or bluish-gray.

6. Does the rock consist predominantly of calcite or dolomite?
   If yes, the rock is carbonatite (Table 5.7).
   If no, the rock is probably either a quartz-poor or feldspathoid-bearing/rich syenitoid (Tables 5.4 and 5.6).
7. What mafic mineral predominates?

If olivine predominates, the rock is peridotite; and if olivine constitutes all or nearly all of the rock, it is the dunite sub-variety of peridotite.
If augite (or other similar-looking pyroxene mineral) predominates and olivine and/or hornblende are common, the rock is pyroxenite.
If hornblende predominates and olivine and/or augite are common, the rock is hornblendite.
If magnetite predominates, it is a magnetitite.
If chromite predominates, it is a chromitite.

8. Does the rock consist predominantly of calcite or dolomite?
   If yes, the rock is carbonatite (Table 5.7).
   If no, go to 9.
9. What is the overall coloration?
   If the rock is light colored, it is either a rhyolite, dacite, phonolite, tephrite, one of the trachytoids, or, less likely, a welded tuff (Tables 5.3-5.4 and 5.6). If phenocrysts are present, try to identify the mineral(s) to further narrow the possibilities (Table 5.9).
   If the rock is dark-gray or black, it is almost certainly either an ordinary or feldspathoid-bearing basalt (Table 5.5). It is definitely basalt if the rock is either vesicular or amygdaloidal.
   If the rock is not obviously one of the above in terms of coloration, it may be an ordinary or feldspathoid-bearing andesite (Table 5.5). Again, look at the phenocrysts, if present, to say more.
10. The rock is obsidian (or, less likely, tachylyte) if it is massive or pumice if it is highly porous (Table 5.8).
11. Are the grains almost all yellowish-green?
    If yes, the rock is dunite (a sub-variety of peridotite), which sometimes has a pseudo-clastic texture (Table 5.7).
    If no, go to 12.
12. What is the predominant size of the grains (clasts) within the rock (Table 5.8)?
    If less than 2 mm, the rock is ash-tuff.
    If between 2 mm and 6.4 cm, the rock is lapilli-tuff.
    If more than 6.4 cm, the rock is either pyroclastic agglomerate (if the clasts are mostly rounded) or pyroclastic breccia (if the clasts are mostly angular).

Table 5.1: Common igneous rock-forming minerals and tips for their macroscopic identification.

FELSIC MINERALS[1]
Quartz • Usually colorless, white, or light- to medium-gray; rarely bluish or other colors. • White streak. • Transparent to translucent with a nonmetallic vitreous to occasionally greasy luster. • No cleavage but well-developed conchoidal fracture. • Brittle with a Mohs hardness of 7. • No secondary alteration. • Phenocrysts occur only in felsic volcanic rocks and are short hexagonal dipyramidal crystals; edges are commonly rounded and embayed due to corrosion. • Almost always occurs as anhedral grains in plutonic rocks. • Fusibility is 7. • Specific gravity is 2.65. • Largely restricted to felsic rocks.
Potassic Feldspar • Usually orthoclase or microcline in plutonic phaneritic rocks, and usually anorthoclase or sanidine in volcanic aphanitic rocks. • Usually pink or red, but occasionally white to light-gray, yellowish-white to yellowish-gray, or brownish-yellow to yellowish-brown; exceptionally green to greenish-blue in microcline for the gemstone known as amazonite, or colorless in sanidine. • White streak. • Translucent to pseudo-opaque with a nonmetallic vitreous to pearly luster. • Marginally prominent cleavage in two directions at a nearly 90° angle. • Brittle with a Mohs hardness of 6-6.5. • No striations on cleavage surfaces, but microcline usually has white to light-gray, roughly parallel perthite lamellae (i.e. irregular stringers or veinlets of albite plagioclase) that are visible on cleavage surfaces. • Twinned crystals are common, especially Carlsbad twins (see Figure 3.4). • Phenocrysts can occur in both volcanic and plutonic rocks, but are commonly euhedral in the former and subhedral in the latter; they are short prismatic or tabular crystals with rectangular outlines in all orientations, and are commonly compositionally zoned with concentric color bands. • Fusibility is 5. • Specific gravity is 2.54-2.57. • Largely restricted to felsic rocks.
Plagioclase Feldspar • Includes, in a solid-solution series, the 'sodic' plagioclases – albite (An 0-10), oligoclase (An 10-30), and andesine (An 30-50); and the 'calcic' plagioclases – labradorite (An 50-70), bytownite (An 70-90), and anorthite (An 90-100); if potassic feldspar is abundant in a rock, the plagioclase is usually sodic, but if it occurs in the same rock with olivine or pyroxene, the plagioclase is always calcic. • Usually white to light-gray for An < 50, and medium- to dark- or bluish-gray for An > 50; rarely colorless, pink, or light-yellow. • White streak. • Translucent to pseudo-opaque with a nonmetallic vitreous to pearly luster. • Marginally prominent cleavage in two directions at a nearly 90° angle. • Brittle with a Mohs hardness of 6-6.5.

[1] See also Tables 3.4 and 3.6 for additional mineral properties.

- Fine, parallel striations, due to polysynthetic twinning (see Figure 3.4), are commonly present on cleavage surfaces.
- Phenocrysts can occur in both volcanic and plutonic rocks, but are commonly euhedral in the former and subhedral in the latter; they are short to long, narrow prismatic crystals with rectangular outlines in all orientations, and are commonly compositionally zoned with concentric color bands.
- Fusibility is 4-4.5.
- Specific gravity is 2.62-2.76.
- Sodic varieties are restricted to felsic and some intermediate rocks, and calcic varieties are restricted to mafic and some intermediate rocks.

All Feldspars: [1] The potassic feldspars together with albite plagioclase are referred to as the **alkali feldspars**. The albite occurs as macroscopic perthite lamellae within microcline; but also as 'microperthite' (i.e. microscopic perthite lamellae) in both microcline and orthoclase. [2] A given rock will usually contain one variety of either potassic or plagioclase feldspar, or both. Only rarely will there be two varieties of potassic feldspar in the same rock, and almost never will two varieties of plagioclase be present. [3] Both potassic and especially plagioclase feldspars are highly susceptible to secondary alteration during late-stage magmatism and weathering; consequently, these minerals are commonly replaced by either clay minerals (giving the feldspar a dull luster), or chlorite and/or epidote (giving the feldspar a greenish color).

Feldspathoid Minerals

- Only occur in the absence of quartz, and one or two varieties of feldspathoid mineral can be present in the same rock.
- There are numerous feldspathoid minerals, all relatively rare, with the most common nepheline (in both phaneritic and aphanitic rocks) followed by leucite (in aphanitic rocks), and cancrinite and sodalite (in phaneritic rocks and commonly in association with nepheline).
- **Nepheline**: usually colorless, white to yellowish-white, or light-gray but occasionally bluish-green, greenish-gray, purplish-brown, or red; transparent to mainly translucent or pseudo-opaque; vitreous to greasy luster; and usually in massive granular aggregates.
- **Leucite**: usually white to light-gray or, less commonly, colorless; translucent to pseudo-opaque; vitreous to dull luster; and usually in trapezohedral or other multi-faceted equidimensional crystals.
- **Cancrinite**: usually white or yellow, but occasionally grayish-green or shades of blue, orange, pink, or red; transparent to translucent; vitreous to greasy or pearly luster.
- **Sodalite**: usually blue, but occasionally white, gray, green, or pink; transparent to mainly translucent; vitreous luster.
- White streak for all varieties.
- No prominent cleavage, but conchoidal fracture for leucite and nepheline.
- Brittle with a Mohs hardness of 5.5-6 for leucite and nepheline.
- These minerals weather more rapidly than the otherwise similar-looking feldspars and, consequently, they can sometimes be recognized on this basis when a rock contains both mineral groups (e.g. on weathered surfaces, the feldspathoid grains are partially replaced by dissolution pits whereas the feldspar grains are not).
- Only nepheline phenocrysts are common and these are hexagonal, short to long prismatic crystals with hexagonal and rectangular outlines perpendicular and parallel, respectively, to the c axis.
- Fusibilities are 4 for nepheline and 7 for leucite.
- Specific gravities for leucite and nepheline are 2.47-2.50 and 2.55-2.66, respectively.
- Can occur in felsic, intermediate, or mafic rocks.

MAFIC MINERALS[1]

Major Varieties

(those occurring in greater abundance and in macroscopically visible grains)

Biotite and **Phlogopite Micas**

- Usually dark-brown or black and rarely dark-green for biotite, and usually pale-yellow or yellowish-brown and rarely greenish- to reddish-brown for phlogopite.
- White to pale-gray streak.
- Transparent to translucent with a submetallic to mainly nonmetallic vitreous or pearly luster.
- Prominent cleavage in one direction with easy separation.
- Flexible and elastic with a Mohs hardness of 2-3.
- When biotite is common, the plagioclase is sodic and potassic feldspar is usually present.
- Other mafic minerals, especially hornblende and magnetite, tend to occur in clumps with biotite.
- Phenocrysts are tabular crystals with hexagonal outlines perpendicular to the c axis; edges are commonly rounded or embayed due to corrosion.
- Weakly attracted to a neodymium magnet.
- Fusibility is 5.
- Specific gravity is 2.70-3.30.
- Biotite is mainly found in felsic to intermediate rocks, but can occur in mafic rocks. Phlogopite is primarily a metamorphic mineral, but sometimes occurs in ultramafic rocks and is commonly associated with leucite in feldspathoid rocks.

Muscovite Mica

- Usually colorless or pale-yellow.
- White streak.
- Transparent to translucent with a nonmetallic vitreous to pearly or silky luster.
- Prominent cleavage in one direction with easy separation.
- Flexible and elastic with a Mohs hardness of 2-3.5.
- Does not occur as phenocrysts.
- Fusibility is 4-5.
- Specific gravity is 2.70-2.90.
- Always occurs with sodic plagioclase and consequently is only found in felsic rocks.

Amphibole Minerals: usually **hornblende** but other similar-appearing varieties are possible

- Usually greenish-black, but occasionally dark-green or dark-brown.
- White to pale-gray or pale-greenish-gray streak.
- Translucent to pseudo-opaque with a nonmetallic vitreous luster.
- Prominent cleavage in two directions at ~56° and ~124° angles.
- Brittle with a Mohs hardness of 5-6 for hornblende.
- Commonly occurs in clumps with biotite and magnetite.
- Phenocrysts are short to mainly long prismatic crystals with diamond-shaped to pseudo-hexagonal and rectangular outlines perpendicular and parallel, respectively, to the c axis.
- Weakly attracted to a neodymium magnet.
- Fusibility is 4 for hornblende.
- Specific gravity is 3.00-3.59 for hornblende.
- Usually only in felsic to intermediate rocks, but can occur in mafic rocks.

Pyroxene Minerals: usually **augite** but other similar-appearing varieties are possible

- Usually dark-green or greenish-black, but occasionally brownish-black.
- White to pale-green streak.
- Translucent to pseudo-opaque with a nonmetallic vitreous luster.
- Marginally prominent cleavage in two directions at a nearly 90° degree angle.
- Brittle with a Mohs hardness of 5.5-6.5 for augite.
- When pyroxene is common, any associated plagioclase is calcic and potassic feldspar is either rare or absent.
- Phenocrysts are short prismatic crystals with pseudo-octagonal and rectangular outlines perpendicular and parallel, respectively, to the c axis.
- Fusibility is 4-4.5 for augite.
- Specific gravity is 3.19-3.56 for augite.
- Usually only in intermediate to mafic rocks, but can occur in felsic rocks.

Olivine

- Usually light- to medium-green or yellowish-green (called **peridot** when gem quality), but occasionally brownish.
- White to pale-gray streak.
- Transparent to translucent with a nonmetallic vitreous luster.
- No cleavage but well-developed conchoidal fracture.
- Brittle with a Mohs hardness of 6.5-7.
- Phenocrysts are usually highly corroded and so appear as rounded granules with roughly circular outlines; only rarely occurs as prismatic or tabular crystals.
- Fayalite (iron-rich olivine) is weakly attracted to a neodymium magnet.
- Fusibility is 7.
- Specific gravity is 3.22-4.39.
- Only occurs in mafic or ultramafic rocks, and never in the same rock with quartz.

Minor Varieties
(those occurring in lesser abundance and in smaller,
but commonly macroscopically visible grains)

Magnetite

- Black whole and streak colors.
- Opaque with a metallic to submetallic luster.
- No characteristic breakage mode.
- Brittle with a Mohs hardness of 5.5-6.5.
- Strongly attracted to neodymium and other weaker magnets.
- Commonly occurs in clumps with biotite and hornblende.
- Fusibility is 5-5.5.
- Specific gravity is 5.17-5.20 g/cc.
- Nearly ubiquitous in crystalline igneous rocks.

Pyrrhotite

- Yellowish-brown to brownish-yellow, darkening when tarnished.
- Dark-gray to black streak.
- Opaque with a metallic luster.
- No characteristic breakage mode.
- Brittle with a Mohs hardness of 3.5-4.5.
- Strongly attracted to a neodymium magnet and, on this basis, is distinguishable from the similar-looking but non-magnetic pyrite.
- Fusibility is 2.5-3.
- Specific gravity is 4.58-4.65 g/cc.
- Largely restricted to mafic rocks.

Note: both magnetite and pyrrhotite are commonly replaced during weathering by reddish hematite or yellowish to brownish goethite/limonite.

Chromite

- Usually black or brownish-black.
- Dark-brown streak.
- Pseudo-opaque to opaque with a metallic to submetallic luster.
- No characteristic breakage mode.
- Brittle with a Mohs hardness of 5.5-6.
- Weakly attracted to a neodymium magnet.
- Fusibility is 7.
- Specific gravity is 4.18-5.50 g/cc.
- Largely restricted to ultramafic rocks.

Ilmenite

- Black to brownish-black whole and streak colors.
- Opaque with a metallic to submetallic luster.
- No characteristic breakage mode.
- Brittle with a Mohs hardness of 5-6.
- Moderately attracted to a neodymium magnet.
- Fusibility is 7.
- Specific gravity is 4.72-4.79 g/cc.
- Largely restricted to mafic and ultramafic rocks.

Table 5.2: Simplified IUGS classification of igneous rocks.

<table>
<tr><td colspan="5">FELSIC TO MAFIC PHANERITIC AND APHANITIC CRYSTALLINE ROCKS
(mafic minerals < 90% of whole rock)
ALKALI FELDSPAR as a volume percent of total (alkali + plagioclase) feldspar[1]
100% — 65% — 50% — 35% — 10% — 0%</td></tr>
<tr><td rowspan="3">QUARTZ as a volume percent of quartz plus total feldspar</td><td>100–60%</td><td colspan="3">HYDROTHERMAL QUARTZ + QUARTZ-RICH GRANITOID phanerites</td></tr>
<tr><td>60–20%</td><td colspan="3">Table 5.3 GRANITOID phanerites & RHYOLITOID + DACITOID aphanites</td></tr>
<tr><td>20–0%</td><td rowspan="2">Tables 5.4
SYENITOID phanerites & TRACHYTOID aphanites</td><td rowspan="2">mafics >10%
GABBROID phanerites & BASALTOID aphanites (An >50)
DIORITOID phanerites & ANDESITOID aphanites (An <50)</td><td rowspan="2">mafics <10%
ANORTHO-SITIC phanerites
Table 5.5</td></tr>
<tr><td rowspan="3">FELDSPATHOIDS as a volume percent of feldsparthoids plus total feldspar</td><td><10%</td></tr>
<tr><td>10–60%</td><td>FELDSPATHOID SYENITOID phanerites & PHONOLITOID aphanites</td><td colspan="2">Table 5.6
FELDSPATHOID DIORITOID + GABBROID phanerites & TEPHRITOID aphanites</td></tr>
<tr><td>>60%</td><td colspan="3">FOIDOLITE phanerites</td></tr>
<tr><td colspan="2">general composition</td><td>light-colored[2] felsic rocks (a.k.a., felsites) with felsic minerals predominating</td><td>intermediate rocks with significant amounts of both felsic and mafic minerals</td><td>dark-colored[3] mafic rocks (a.k.a., mafites) with mafic minerals predominating)</td></tr>
</table>

<table>
<tr><td>ULTRAMAFIC PHANERITIC CRYSTALLINE ROCKS
(mafic minerals > 90% of whole rock)</td><td>Table 5.7</td></tr>
</table>

<table>
<tr><td>GLASSY ROCKS and PYROCLASTIC ROCKS</td><td>Table 5.8</td></tr>
</table>

[1] Alkali feldspars include the potassic feldspars plus albite plagioclase.
[2] White, light- to medium-gray, pink, red, yellow, light-brown, and light- to medium-green.
[3] Dark-gray, dark-green, brownish-black, and black.

Table 5.3: Detailed IUGS classification of igneous quartzose and granitoid phanerites, and their rhyolitoid and dacitoid aphanite equivalents.

<table>
<tr><td rowspan="10">QUARTZ as a volume percent of quartz plus total feldspar</td><td>>90%</td><td colspan="3">QUARTZOLITE</td><td>Milky white, translucent vein quartz (a.k.a. hydrothermal quartz).</td></tr>
<tr><td>90-60%</td><td colspan="4">quartz-rich granitoids (extremely rare)</td></tr>
<tr><td rowspan="8">60-20%</td><td rowspan="8">ALKALI FELDSPAR as a volume percent of total (alkali plus plagioclase) feldspar</td><td colspan="3">Phaneritic Granitoid Rocks</td></tr>
<tr><td>100-90%</td><td>ALKALI-FELDSPAR GRANITE
-
CI = 0-20</td><td rowspan="4">• Overall coloration: light- to medium-gray, pink to nearly red, or brownish-yellow. Upon closer examination, these rocks show a mottling of light felsic and dark mafic minerals. Usually only granites and some granodiorites, due to abundant potassic feldspar, will be pinkish or reddish. Exceptions can occur: e.g. if biotite or hornblende are abnormally abundant, the rock can be dark-gray to nearly black; and if the feldspar is altered to chlorite or epidote, the rock can acquire a greenish cast.
• Feldspar minerals: alkali – either orthoclase or microcline with the latter commonly perthetic; and plagioclase – usually oligoclase but occasionally andesine.
• Phenocrysts: mostly orthoclase and perthitic microcline in granites, and mostly plagioclase in granodiorite and tonalite.
• Mafic minerals: mainly biotite and hornblende plus occasionally muscovite (in granites only) and augite (or other pyroxenes). The first two minerals plus magnetite commonly occur together in clumps.
• Grain form: the feldspar and mafic minerals commonly show some degree of crystal form in the groundmass but quartz shows none.
• Special sub-varieties: granitoids with a grain size of ~1-2 mm are called aplites. In graphic granite the quartz occurs as long, nearly parallel, wedge-shaped prisms within the larger alkali feldspar grains. Miarolitic granitoids contain open geode-like cavities lined with crystals of one or more mineral varieties.</td></tr>
<tr><td>90-35%</td><td>GRANITE
-
CI = 5-20
SG = 2.5-2.8 g/cc</td></tr>
<tr><td>35-10%</td><td>GRANODIORITE
-
CI = 5-25
SG = 2.7-2.8 g/cc</td></tr>
<tr><td>10-0%</td><td>TONALITE
-
CI = 10-40
SG = ~2.76 g/cc</td></tr>
<tr><td colspan="3">Aphanitic Rhyolitoid and Dacitoid Rocks</td></tr>
<tr><td>100-35%</td><td>RHYOLITE</td><td rowspan="2">• Overall coloration: light- to medium-gray, pink to nearly red, or brownish-yellow with rhyolite more likely to be pinkish or reddish. Rhyolite is occasionally rich in magnetite, which can give it a nearly black color. A near-glassy translucency is common in these rocks as is also flow-banding marked by color differences and/or parallel alignment of phenocrysts. The coloration of aphanitic rocks is more uniform than in phaneritic rocks of the same composition.
• Feldspar minerals: alkali – usually either orthoclase or sanidine, rarely anorthoclase; and plagioclase – either oligoclase or andesine.
• Mafic minerals: mainly biotite with subordinate augite (or other pyroxenes) and hornblende (or other amphiboles).
• If porphyritic, see Table 5.9 for the likely mineralogy of the phenocrysts.</td></tr>
<tr><td>35-0%</td><td>DACITE</td></tr>
</table>

Note 1: the relative global abundance of rock varieties is indicated by the rock name font: '**ABUNDANT**', 'COMMON', and 'rare'.

Note 2: CI = color index, and SG = specific gravity.

Table 5.4: Detailed IUGS classification of igneous syenitoid phanerites, and their trachytoid aphanite equivalents.

<table>
<tr><td rowspan="7">QUARTZ as a volume percent of quartz plus total feldspar</td><td rowspan="4">20-5%</td><td rowspan="10">ALKALI FELDSPAR as a volume percent of total (alkali plus plagioclase) feldspar</td><th colspan="3">Phaneritic Syenitoid Rocks</th></tr>
<tr><td>100-90%</td><td>alkali-feldspar quartz syenite
–
CI = 5-25</td><td rowspan="9">• Overall coloration: usually pink to nearly red or brownish-yellow, and less commonly light-gray or light-yellow. Upon closer examination, these rocks show a mottling of light felsic and dark mafic minerals.
• Feldspar minerals: alkali – either orthoclase or perthetic microcline; and plagioclase – either oligoclase or andesine.
• Mafic minerals: mainly biotite and hornblende (or other amphiboles) with subordinate augite (or other pyroxenes) and olivine.
• 'Foid' is a placeholder for the name of the predominant feldspathoid mineral: e.g. nepheline-bearing syenite.</td></tr>
<tr><td>90-65%</td><td>quartz syenite
–
CI = 5-30</td></tr>
<tr><td>65-35%</td><td>quartz monzonite
–
CI = 10-35
SG = ~2.64 g/cc</td></tr>
<tr><td rowspan="3">5-0%</td><td>100-90%</td><td>alkali-feldspar syenite
–
CI = 0-25</td></tr>
<tr><td>90-65%</td><td>syenite
–
CI = 10-35</td></tr>
<tr><td>65-35%</td><td>monzonite
–
CI = 15-45</td></tr>
<tr><td rowspan="3">FELDSPATHOIDS as a volume percent of feldspathoids plus total feldspar

(no quartz)</td><td rowspan="3">0-10%</td><td>100-90%</td><td>foid-bearing alkali-feldspar syenite
–
CI =0-25</td></tr>
<tr><td>90-65%</td><td>foid-bearing syenite
–
CI =10-35%</td></tr>
<tr><td>65-35%</td><td>foid-bearing monzonite
–
CI = 15-45%</td></tr>
<tr><td rowspan="5">QUARTZ as a volume percent of quartz plus total feldspar</td><td rowspan="3">20-5%</td><td rowspan="7">ALKALI FELDSPAR as a volume percent of total (alkali plus plagioclase) feldspar</td><th colspan="3">Aphanitic Trachytoid Rocks</th></tr>
<tr><td>100-65%</td><td>quartz trachyte</td><td rowspan="6">• Overall coloration and 'foid' placeholder: the same as the phaneritic (syenitoid) rocks above.
• Feldspar minerals: anorthoclase or sanidine; and plagioclase – either oligoclase or andesine.
• Other: flow-banding is common in the aphanitic rocks with this marked by color differences and/or parallel alignment of phenocrysts.
• If porphyritic, see Table 5.9 for the likely mineralogy of the phenocrysts.</td></tr>
<tr><td>65-35%</td><td>quartz latite</td></tr>
<tr><td rowspan="2">5-0%</td><td>100-65%</td><td>trachyte</td></tr>
<tr><td>65-35%</td><td>latite</td></tr>
<tr><td rowspan="2">FELDSPATHOIDS as a volume percent of feldspathoids plus total feldspar

(no quartz)</td><td rowspan="2">0-10%</td><td>100-65%</td><td>foid-bearing trachyte</td></tr>
<tr><td>65-35%</td><td>foid-bearing latite</td></tr>
</table>

Note 1: the relative global abundance of rock varieties is indicated by the rock name font: '**ABUNDANT**', 'COMMON', and 'rare'.
Note 2: CI = color index, and SG = specific gravity.

Table 5.5: Detailed IUGS classification of igneous dioritoid, gabbroid and anorthositic phanerites, and their andesitoid and basaltoid aphanite equivalents.

			Phaneritic *Dioritoid, Gabbroid* and *Anorthositic* Rocks				
				Plagioclase An <50	Plagioclase An >50		
QUARTZ as a volume percent of quartz plus total feldspar	20-5%	ALKALI FELDSPAR as a volume percent of total (alkali plus plagioclase) feldspar	35-10%	quartz monzodiorite – CI = 15-40	quartz monzogabbro – CI = 20-50		• Overall coloration: medium-gray to greenish-gray for dioritoids and dark-greenish- or bluish-gray to black for gabbroids. Upon closer examination, especially for the dioritoids, these rocks appear to be speckled or mottled with light-colored plagioclase and dark-colored mafic minerals. The anorthosites, which do not have an aphanitic equivalent, are lighter colored, and are usually medium- to dark-gray or, more often, bluish-gray, and commonly exhibit labradorescence from labradorite plagioclase. • Feldspar minerals: alkali – orthoclase; and plagioclase – oligoclase to mainly andesine to anorthite. Although the An content of the plagioclase in anorthosite and quartz anorthosite are usually over 50%, it is occasionally as low as 40%. • Mafic minerals: mainly hornblende in dioritoids and augite in gabbroids plus lesser amounts of biotite, olivine, and others (mainly magnetite, pyrrhotite, and/or ilmenite). • 'Foid' is a placeholder for the name of the predominant feldspathoid mineral: e.g. nepheline-bearing diorite. • Special sub-variety: gabbroids with a grain size of ~1-2 mm are called **diabase** in the US and **dolerite** in the UK.
			10-0%	quartz diorite – CI = 20-45 SG = 2.7-3.0 g/cc	quartz gabbro *(mafics >10% of whole rock)* – CI = 25-55	quartz anorthosite *(mafics <10% of whole rock)*	
	5-0%		35-10%	monzodiorite – CI = 20-50	monzogabbro – CI = 25-60		
			10-0%	**DIORITE** – CI = 25-50 SG = 2.7-3.0 g/cc	**GABBRO** *(mafics >10% of whole rock)* – CI = 35-65 SG = 2.7-3.1 g/cc	ANORTHOSITE *(mafics <10% of whole rock)* – CI = 0-10 SG = 2.6-2.9 g/cc	
FELDSPATHOIDS as a volume percent of feldspathoids plus total feldspar (no quartz)	0-10%		35-10%	foid-bearing monzodiorite/monzogabbro			
			10-0%	foid-bearing diorite/gabbro			

<table>
<tr><td>QUARTZ as a volume percent of quartz plus total feldspar</td><td>20-0%</td><td rowspan="4">ALKALI FELDSPAR as a volume percent of total (alkali plus plagioclase) feldspar</td><td colspan="4">Aphanitic Andesitoid and Basaltoid Rocks</td></tr>
<tr><td rowspan="3">FELDSPATHOIDS as a volume percent of feldspathoids plus total feldspar

(no quartz)</td><td rowspan="3">0-10%</td><td rowspan="3">35-0%</td><td>Plag. An <50</td><td>Plag. An >50</td><td rowspan="3">• Overall coloration: medium-gray to greenish-gray for andesite and dark-gray to mainly black for basalt. Both rock types, especially basalt, can be greenish when secondary chlorite and/or epidote is abundant. Such rocks are gradational with metamorphic greenstone.
• Flow banding as in the felsic aphanites (Tables 5.3-5.4) is common in light-colored andesite but rare in dark-colored andesite and basalt.
• Mafic minerals are the same as in the dioritoids and gabbroids, but the proportions commonly differ: i.e. hornblende ≥ biotite >> pyroxene for andesite, and pyroxene >> hornblende > biotite for basalt.
• Both andesite and, especially, basalt can be vesicular or amygdaloidal.
• 'Foid' is a placeholder for the name of the predominant feldspathoid mineral: e.g. nepheline-bearing andesite.
• If porphyritic, see Table 5.9 for the likely mineralogy of the phenocrysts.</td></tr>
<tr><td>ANDESITE</td><td>BASALT</td></tr>
<tr><td colspan="2">foid-bearing andesite/basalt</td></tr>
</table>

Note 1: the relative global abundance of rock varieties is indicated by the rock name font: '**ABUNDANT**', 'COMMON', and 'rare'.

Note 2: CI = color index, and SG = specific gravity.

Table 5.6: Detailed IUGS classification of the igneous feldspathoid-rich syenitoid, dioritoid and gabbroid phanerites, and their phonolitoid and tephritoid aphanite equivalents.

FELDSPATHOIDS as a volume percent of feldspathoids plus total feldspar (no quartz)		ALKALI FELDSPAR as a volume percent of total (alkali plus plagioclase) feldspar		Rock	Description
			Phaneritic *Syenitoid, Dioritoid* and *Gabbroid* Rocks		
	10-60%		100-90%	'foid' syenite – CI = 0-30	All these rocks superficially resemble the similarly-named syenitoid, dioritoid, and gabbroid rocks in Tables 5.4-5.5. They have the same feldspar and mafic mineral contents, but the latter are mainly other dark varieties of amphiboles and pyroxenes, the so-called 'pyribole' minerals. Quartz has been replaced by one or more of the feldspathoid minerals. The two most common of these - leucite and, especially, nepheline - are not macroscopically distinguishable from quartz and the feldspars unless the feldspathoid minerals exhibit distinctive colors, such as green nepheline, yellow cancrinite, or blue sodalite. Nepheline and leucite weather more rapidly than the feldspars and so can be distinguished from them in weathered specimens where it appears as surface depressions or pits. 'Foid' is a placeholder for the name of the predominant feldspathoid mineral: e.g. nepheline syenite.
	10-60%		90-50%	'foid' monzosyenite – CI = 15-45	
	10-60%		50-10%	'foid' monzodiorite/ monzogabbro – CI = 20-60	
	10-60%		10-0%	'foid' diorite/gabbro – CI = 30-70	
			Aphanitic *Phonolitoid* and *Tephritoid* Rocks		
	10-60%		100-50%	PHONOLITE	These rocks have essentially the same coloration and mineralogy as the phaneritic equivalents above, although the phonolites are typically medium-grayish-green.
	10-60%		50-0%	tephrite	
			Phaneritic *Foidolitoid* Rocks		
	60-100%		100-0%	foidolite – CI = 0-90	This category is more theoretical than real because this rock type has not yet been found in nature.

Note 1: the relative global abundance of rock varieties is indicated by the rock name font: '**ABUNDANT**', 'COMMON', and 'rare'.
Note 2: CI = color index.

Table 5.7: IUGS classification of igneous ultramafic phanerites.

ULTRAMAFIC ROCKS (mafic minerals >90% of whole rock)	
PERIDOTITE	• Over 40% olivine with the remainder mainly other mafic minerals (primarily pyroxene, but also amphibole and/or phlogopite) plus minor calcic plagioclase. • Dark-green to black color, sometimes with a speckling of white or pale-gray plagioclase grains. • Mafic minerals are commonly partially or largely replaced by serpentine minerals (i.e. the rock has been altered to metamorphic serpentinite). • Specific gravity = 3.15-3.28 g/cc.
dunite (a sub-variety of peridotite)	• Over 90% olivine with the remainder other mafic minerals (primarily chromite, ilmenite, magnetite, and/or pyrrhotite). • Light yellowish-green color (can become brownish upon weathering). • Although a phaneritic crystalline rock, dunite's texture sometimes superficially resembles the clastic texture of a sedimentary rock because of the corrosive rounding of the olivine crystals. • Specific gravity = 3.20-3.31 g/cc.
pyroxenite	• Less than 40% olivine with the remainder mainly pyroxene. • Greenish-black, black, or brown color. • Pyroxene is commonly partially or largely replaced by secondary hornblende (i.e. the rock has been altered to hornblendite). • Specific gravity = 3.10-3.32 g/cc.
hornblendite	• Less than 40% olivine with the remainder mainly hornblende. • Black to greenish-black color. • May originally have been pyroxenite.
magnetitite	• Over 90% mafic minerals, mainly magnetite. • Black color. • With increasing magnetite content, the specific gravity approaches that of the mineral (5.17-5.20 g/cc). • Strongly magnetic with some specimens acting like magnets (i.e. 'lodestones').
chromitite	• Over 90% mafic minerals, mainly chromite. • Black to brownish-black color. • With increasing chromite content, the specific gravity approaches that of the mineral (4.18-5.50 g/cc).
carbonatite	• Over 90% carbonate mineral – usually calcite and occasionally dolomite. • Light-gray to yellowish-white color. • Although calcite and dolomite are not mafic minerals, carbonatite is grouped with ultramafic rocks in the IUGS classification because, like them, it has a very low silica content.

Note: the relative global abundance of rock varieties is indicated by the rock name font: '**ABUNDANT**', 'COMMON', and 'rare'.

Table 5.8: IUGS classification of glassy and pyroclastic rocks.

<table>
<tr><th colspan="4">GLASSY ROCKS (glass >80% of whole rock)</th></tr>
<tr><td rowspan="3">ALKALI FELDSPAR as a volume percent of total (alkali plus plagioclase) feldspar</td><td rowspan="2">100-10%</td><td>OBSIDIAN
(massive glass)</td><td>• Composition: usually rhyolitic or dacitic, but occasionally andesitic, latitic, trachytic, or phonolitic.
• Appearance: usually dark-gray or black in color, but occasionally red, brown, green, or nearly colorless with two or more colors sometimes occurring as deformed streaks; transparent to mainly translucent with a vitreous luster and an especially well-developed conchoidal fracture.
• Porphyritic rocks with a largely glassy, obsidian-like, groundmass can occur and these are called vitrophyres.
• Geologically ancient obsidian (i.e. over about 100 million years old) alters through devitrification and/or hydration to pitchstone, which resembles obsidian except for its dull resinous or pitch-like luster.
• Specific gravity is 2.3-2.4 g/cc.</td></tr>
<tr><td>PUMICE
(frothy glass)</td><td>• Composition: same as obsidian.
• Appearance: usually nearly white, light-gray or light-yellow in color, but occasionally brown or red. Transparent to mainly translucent with a silky luster.
• Highly porous; most varieties float on water.</td></tr>
<tr><td>35- 0%</td><td>tachylyte
(massive glass)</td><td>• Composition: usually basaltic, but occasionally andesitic.
• Usually occurs as clots and stringers within andesite and especially basalt.
• Appearance: greenish-black to black in color. Translucent to pseudo-opaque with a greasy luster and a conchoidal fracture.</td></tr>
<tr><th colspan="4">PYROCLASTIC ROCKS</th></tr>
<tr><td colspan="2">ASH-TUFF</td><td colspan="2">• Indurated pyroclastic 'ash' (mainly glass shards and mineral fragments; less commonly rock fragments) with grain sizes mostly smaller than 2 mm.
• Can resemble some sedimentary sandstones but has rougher surfaces due to the usually abundant glass shards.
• Commonly shows fine color-banding due to either flowage during deposition or sequential sedimentation.
• When grains are fused together, as in a welded tuff, the rock is gradational with obsidian and can superficially resemble sedimentary chert.</td></tr>
<tr><td colspan="2">LAPILLI-TUFF</td><td colspan="2">• Indurated pyroclastic 'lapilli' (mainly pumice and other rock fragments) with grain sizes mostly between 2 mm and 6.4 cm.
• Can resemble some sedimentary conglomerates but the pumice clasts indicate an igneous origin.
• Occasionally shows fine color-banding.</td></tr>
<tr><td colspan="2">pyroclastic agglomerate and breccia</td><td colspan="2">• Indurated pyroclastic debris (mainly rock fragments – blocks and clots of hardened lava – a.k.a. bombs) with grain sizes mostly larger than 6.4 cm and clasts either rounded (agglomerate) or angular (breccia).
• Can resemble some sedimentary conglomerates and breccias.</td></tr>
</table>

Note: the relative global abundance of rock varieties is indicated by the rock name font: '**ABUNDANT**', 'COMMON', and 'rare'.

Table 5.9: Typical phenocrysts assemblages in aphanitic (volcanic) igneous rocks.

PHENOCRYSTS[1]		APHANITIC ROCK TYPES						
		RHYOLITE	DACITE	LATITE	TRACHYTE	ANDESITE	BASALT	PHONOLITE
felsic	quartz	▮	■	—	—	—	—	▮[2]
	potassic feldspar	▮	▮	■	▮	—	—	▮
	plagioclase	■	▮	▮	—	▮	▮	—
mafic	biotite	■	■	■	■	■	—	—
	hornblende	■	■	■	■	■	■	—
	augite or other pyroxene	■	■	■	■	■	■	■
	olivine	—	—	■	—	■	■	—

[1] Abundance codes: █ or ▮ = often present, ■ = occasionally present, and — = very rarely or never present. When naming a porphyritic aphanite, add the name of the principal phenocryst mineral (or minerals) as an adjective (e.g. quartz rhyolite porphyry or hornblende-olivine basalt porphyry).
[2] Has nepheline and/or leucite phenocrysts instead of quartz.

5.3 Sedimentary Rocks

5.3.1 Description and Classification

As with igneous rocks, the classification of sedimentary rocks is based on both composition and texture. The first distinction to be made is between the 'silicic' rocks that consist principally of silicate minerals (i.e. clay, feldspar, and especially quartz; Table 5.12) and the rocks dominated by other ('non-silicic') mineralogies (i.e. primarily calcite, but also dolomite, halite, anhydrite, gypsum, organic carbon, iron oxides, and phosphates; Table 5.13). Most of the former have a clastic texture (Figure 5.5) and so are widely referred to as 'siliciclastic rocks'.

Unlike for igneous rocks, those of sedimentary origin have not been defined within a single comprehensive classification. Separate and largely unrelated classification schemes have evolved independently for the three most important sedimentary rock groups: sandy siliciclastics (i.e. sandstones), muddy siliciclastics (i.e. mudrocks), and carbonates (i.e. limestones and dolostones). The other groups of sedimentary rocks – gravelly (i.e. gravelstones), evaporitic, carbonaceous, ferruginous, and phosphatic – each have their own idiosyncratic classifications. All these schemes rely on both macroscopic and microscopic rock properties, with the latter determined by thin-section petrography. To identify sedimentary rocks by strictly macroscopic means, however, it is necessary to synthesize and simplify the various formal classifications, and this has been done in Tables 5.12 and 5.13. Specific gravities are not available for all the sedimentary rocks in these tables, but those reported come from Clarke (1966: 4-4).

Sedimentary rocks are mineralogically simple by comparison to igneous and metamorphic rocks. The weathering and erosional processes that produce the detrital sediments for the clastic rocks remove the physically and chemically unstable minerals, leaving the feldspars, and especially quartz, to make up the bulk of the sandy siliciclastic rocks. The post-depositional diagenetic processes described in section 4.3.2 can add additional minerals in the form of cement (e.g. calcite or iron oxides), or replace the chemically unstable minerals (e.g. clay after feldspar and iron oxides after hornblende).

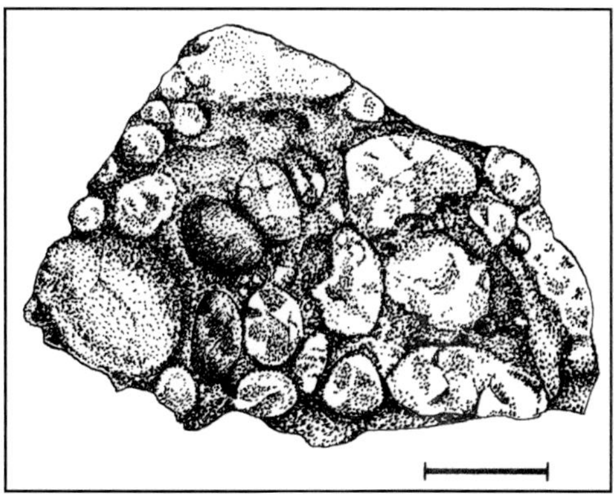

Figure 5.5: A typical example of a clastic rock (conglomerate). From Figure 4 in Raymond (1984) with permission of Waveland Press, Inc. Scale bar = 2 cm.

The Udden-Wentworth grain-size scale has been widely adopted for describing textures in siliciclastic and some other sedimentary rocks (Table 5.10). There are, however, other grain-size scales in common use around the world. Nevertheless, the grain-size definitions for textural terms – e.g. 'clay', 'silt', 'sand', 'pebble', 'cobble', and 'boulder' – are similar, but because there can be small differences among the scales, it is important to always specify which one you are working with. If you prefer a different scale to the Udden-Wentworth, you can still use Tables 5.11-5.12 by simply 'shifting' the grain-size boundaries between the named rock

Table 5.10: Udden-Wentworth grain-size scale for sedimentary rocks.

<table>
<tr><th colspan="3">UDDEN-WENTWORTH SCALE</th></tr>
<tr><th>Millimeters</th><th colspan="2">Size Class</th></tr>
<tr><td>> 256</td><td>boulder</td><td rowspan="4">gravel</td></tr>
<tr><td>64-256</td><td>cobble</td></tr>
<tr><td>4-64</td><td>pebble</td></tr>
<tr><td>2-4</td><td>granule</td></tr>
<tr><td>1-2</td><td>very coarse sand</td><td rowspan="5">sand</td></tr>
<tr><td>0.5-1</td><td>coarse sand</td></tr>
<tr><td>0.25-0.5</td><td>medium sand</td></tr>
<tr><td>0.125-0.25</td><td>fine sand</td></tr>
<tr><td>0.0625-0.125</td><td>very fine sand</td></tr>
<tr><td>0.0039-0.0625</td><td>silt</td><td rowspan="2">mud</td></tr>
<tr><td>< 0.0039</td><td>clay</td></tr>
</table>

varieties. To measure grain size, place a ruler or other scale with millimeter divisions on a specimen and, under a hand-lens, measure the predominant size. The relative amounts of different minerals and matrix will also need to be determined and these can be estimated using the visual comparator in Figure 5.4.

An important determinative property for the coarser-grained siliciclastic rocks is the distinction between 'grain-supported' and 'matrix-supported' textures. In the former, the gravel clasts or sand grains impinge upon each other to create a self-supporting framework whereas in the latter, the gravel clasts and sand grains are largely disconnected and so 'float' in a matrix of finer-grained sediment (usually sand in gravelly rocks, and usually silt and/or clay in sandy rocks). On rare occasions, elongated or tabular gravel clasts will show a preferred orientation due either to the effects of water flow or compaction. For the finer-grained siliciclastic rocks, a critical property is the presence or absence of 'fissility', which is the tendency of a rock to easily split into thin, sub-parallel flakes or sheets (Figure 5.6). Confusingly, but unavoidably, for this same group of rocks, the term 'clay' can refer to either a group of silicate minerals (see Tables 3.6 and 5.11) or the finest size fraction in the Udden-Wentworth scale (Table 5.10). This double meaning comes from the fact that nearly all clay-size sediment grains are composed of clay minerals, with most of the rest quartz. Distinguishing the relative amounts of silt- and clay-size grains in a rock cannot be done visually, but a rough idea of the proportions can be obtained from the 'chew test'. When nipped between the teeth, a mudrock consists of more than one-third silt if it feels gritty, and less than this amount (i.e. over two-thirds clay) if it feels smooth.

Some limestones have neither a clastic nor crystalline texture but, rather, an 'organic framework' (e.g. reef limestone). In these cases, the rock components 'grew' together to

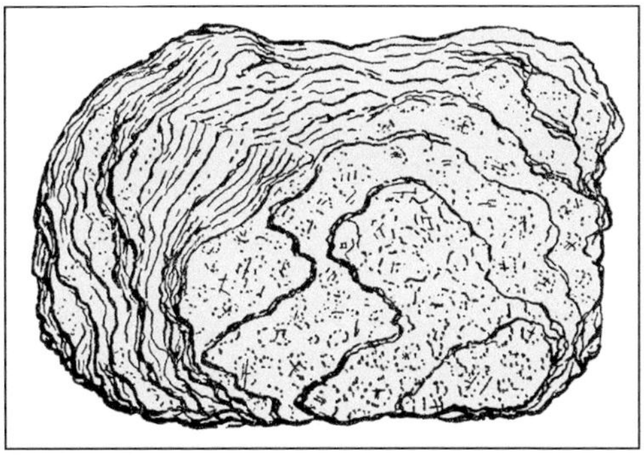

Figure 5.6: Example of a mudrock (shale) displaying good fissility. From Figure 9-3I in Spock (1962) with permission of Harper and Row Publishers.

form a rigid 'reef' fixed to the seabed. Coral reef frameworks are the best known, but other colonial marine invertebrates can also create these structures. On the scale of a hand specimen, however, a reef limestone will be indistinguishable from a fossiliferous limestone. Carbonaceous rocks (coals) have a massive, non-granular texture that is best described as 'amorphous'. Such a texture cannot be distinguished from the finest-grained clastic and crystalline textures, but carbonaceous rocks are easily identified from their other properties (e.g. color, specific gravity, and reaction to flames; see Table 5.13).

Perhaps the most interesting aspect of sedimentary rocks is the fossils they sometimes contain. The skeletal remains of marine invertebrate animals occur occasionally in mudrock but are especially common in limestone (Figures 5.7-5.8). Such rocks can be described as 'fossiliferous' or, equivalently, 'bioclastic'. Table 5.14 summarizes the main macroscopic characteristics of each group of marine invertebrates. It is these fossils that are usually encountered in rocks and among them it is the two main varieties of molluscs that are the most common (i.e. clams/pelecypods and snails/gastropods). Limestones rich in mollusc shells (e.g. Figure 5.8) are often referred to as 'coquina'. The remains of terrestrial invertebrates are exceedingly rare, but fossilized plants are somewhat more common. These include leaf impressions (carbonaceous films) in coal and mudrock, and petrified (silicified) tree limbs and trunks in sandstone. The brownish, phosphatic (but sometimes silicified) bones and teeth of marine

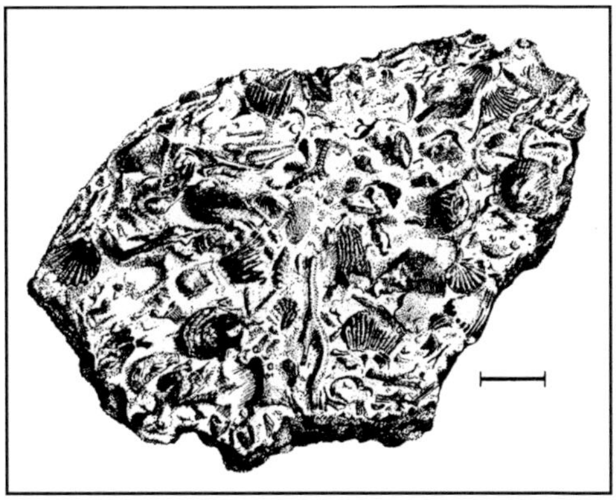

Figure 5.7: Example of fossiliferous (a.k.a. bioclastic) limestone with bryozoan twigs and brachiopod shells. From Figure 38B in Raymond (1984) with permission of Waveland Press, Inc. Scale bar = 2 cm.

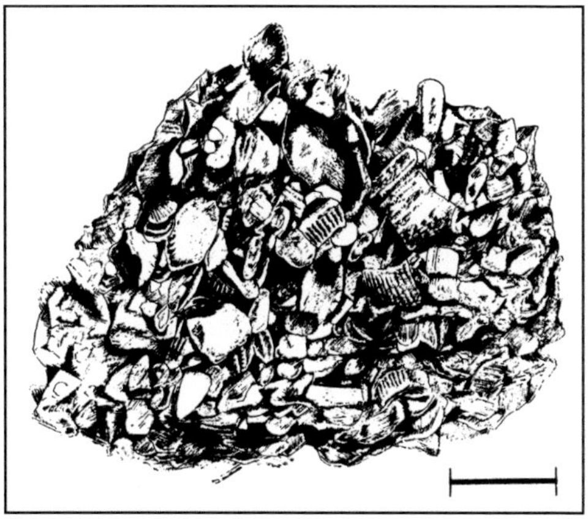

Figure 5.8: Example of coquina or coquinoid (a.k.a. bioclastic or fossiliferous) limestone with pelecypod shells. From Figure 38A in Raymond (1984) with permission of Waveland Press, Inc. Scale bar = 2 cm.

and terrestrial vertebrate animals are the rarest of all fossils and are largely restricted to mudrock and limestone.

5.3.2 *Identification Key*

The following key, when used together with Tables 5.11-5.13, will usually allow you to identify a specific rock variety for an unknown sedimentary specimen.

1. Are the grains macroscopically discernable? (see section 4.7 for a definition)
   If yes, go to 2.
   If no, go to 21.
2. What kind of texture does the rock have?
   If clastic (usually friable to some degree), go to 3.
   If crystalline (usually not friable), go to 18.
   If amorphous and bituminous, go to 23.
   If organic framework, the rock is reef limestone.
3. What is the predominant grain size?
   If over 2.00 mm (gravel size), go to 4.
   If between 0.062 and 2.00 mm, go to 6.
   If less than 0.062 mm, go to 10.
4. What is the composition of the gravel-size clasts?
   If predominantly silicate mineral and/or rock fragments, go to 5.
   If predominantly hematite and/or other iron oxide minerals, the rock is ironstone (Table 5.13).
   If predominantly phosphatic minerals, the rock is phosphorite (Table 5.13).
   If predominantly calcareous invertebrate fossils, the rock is fossiliferous or coquinoid limestone (Table 5.13). Determine the fossil type to further describe the rock variety (Table 5.14).
   If predominantly plant fossils and bituminous, the rock is lignite (Table 5.13).
5. Are the gravel clasts predominantly rounded or angular, and is the rock grain- or matrix-supported?
   If grain supported with rounded clasts, the rock is conglomerate (Table 5.12).
   If grain supported with angular clasts, the rock is breccia (Table 5.12).
   If matrix supported with clasts of any form, the rock is diamictite (Table 5.12).
6. Does the rock consist predominantly of silicate mineral and/or rock fragment grains?
   If yes, go to 7.
   If no, go to 8.
7. Among just the sand grains, what are the relative amounts of quartz, feldspar, and rock fragments in terms of total sand grains, and what is the relative amount of silt/clay matrix in terms of the whole rock?
   If sand grains are over 90% quartz with less than 15% matrix, the rock is quartz sandstone (Table 5.12).
   If sand grains are less than 90% quartz with more feldspar than rock fragments and less than 15% matrix, the rock is arkosic sandstone (Table 5.12).
   If sand grains are less than 90% quartz with more rock fragments than feldspar and less than 15% matrix, the rock is lithic sandstone (Table 5.12).
   If the matrix is over 15% irrespective of grain types, the rock is graywacke sandstone (Table

5.12).

8. What is the predominant composition of the rock?
   If calcite, go to 9.
   If hematite and/or other iron oxide minerals, the rock is ironstone (Table 5.13).
   If phosphatic minerals, the rock is phosphorite (Table 5.13).
   If plant fossils and bituminous, the rock is lignite (Table 5.13).
9. Does the rock consist predominantly of ooliths (sand-size spherical or ovoid grans) or fossil fragments?
   If ooliths, the rock is oolitic limestone (Table 5.13).
   If fossils, the rock is fossiliferous limestone (Table 5.13).
10. What is the predominant composition of the grains?
    If calcite or dolomite, go to 11.
    If silicate minerals, go to 15.
    If hematite and/or other iron oxide minerals, the rock is ironstone (Table 5.13)
    If phosphatic minerals, the rock is phosphorite (Table 5.13).
11. Is the Mohs hardness of the rock less than 2.5?
    If yes, the rock is chalk (Table 5.13).
    If no, go to 12.
12. Is the rock highly porous?
    If yes, the rock is calcareous tufa, a variety of travertine (Table 5.13).
    If no, go to 13.
13. Is the rock translucent and banded?
    If yes, the rock is calcareous sinter, a variety of travertine (Table 5.13).
    If no, the rock is either micritic limestone or crystalline limestone (Table 5.13).
14. What is the rock's color?
    If white or light-gray, it is probably diatomite, but possibly radiolarite or spiculite (Table 5.12). The opaline fossils in these rocks may be visible under a hand-lens and, if so, these will further identify the rock.
    If any other color, especially shades of gray or green, go to 15.
15. Is the rock a mixture of both abundant calcite and silicic silt and/or clay grains?
    If yes, it is marlstone (Table 5.13).
    If no, go to 16.
16. Is the rock fissile?
    If yes, it is the shale sub-variety of mudrock (Table 5.12).
    If no, go to 17.
17. What are the relative amounts of silt- and clay-size grains? Note that the 'chew test' can help to determine these amounts.
    If silt greatly predominates, the rock is the siltstone sub-variety of mudrock (Table 5.12).
    If both silt and clay are abundant or clay predominates, the rock is the mudstone sub-variety of mudrock (Table 5.12).
18. What is the rock's predominant composition?
    If calcite, go to 19.
    If dolomite, the rock is dolostone (Table 5.13).
    If hematite and/or other iron oxide minerals, the rock is ironstone (Table 5.13).
    If phosphatic minerals, the rock is phosphorite (Table 5.13).
    If halite, it is rock salt (Table 5.13).
    If anhydrite, it is rock anhydrite (Table 5.13).

If gypsum, it is rock gypsum (Table 5.13).

19. Is the rock dense (non-porous)?
    If yes, go to 20.
    If no, it is calcareous tufa, a variety of travertine (Table 5.13).
20. Is the rock translucent and banded?
    If yes, the rock is calcareous sinter, a variety of travertine (Table 5.13).
    If no, the rock is crystalline limestone (Table 5.13).
21. Can the rock be scratched with a steel knife blade?
    If yes, go to 22.
    If no, the rock is almost certainly chert, especially if it is dense (non-porous). If highly porous, it is siliceous sinter (Table 5.12).
22. What is the rock's predominant composition?
    If silicate minerals with or without abundant calcite, go to 14.
    If calcite, go to 11.
    If dolomite, the rock is dolostone (Table 5.13).
    If hematite and/or other iron oxide minerals, the rock is ironstone (Table 5.13).
    If phosphatic minerals, the rock is phosphorite (Table 5.13).
    If halite, it is rock salt (Table 5.13).
    If anhydrite, it is rock anhydrite (Table 5.13).
    If gypsum, it is rock gypsum (Table 5.13)
    If bituminous (with a brown, brownish-black, or black color), go to 23.
23. Is the rock soft (Mohs <2) and friable?
    If yes, the rock is normal lignite (Table 5.13).
    If no, go to 24.
24. Does the rock leave a black, powdery residue on the fingers?
    If yes, the rock is bituminous coal (Table 5.13).
    If no, the rock is the jet variety of lignite (Table 5.13).

Table 5.11: Common sedimentary rock-forming minerals and tips for their macroscopic identification.

SILICATE MINERALS
Quartz (see Table 5.1)
Chalcedony (a.k.a. Chalcedonic Quartz) • Usually colorless, gray or white, but when impure can be shades of orange, red, brown, yellow, blue, or green. • White streak. • Translucent to opaque with a nonmetallic waxy or dull luster. • No cleavage, but breaks by conchoidal fracture. • Microscopically fibrous whereas normal quartz is granular. • Brittle with a Mohs hardness of 6.5-7. • Commonly fluoresces pale-yellow or green in short-wavelength UV light. • Fusibility is 7. • Specific gravity is 2.50-2.67. • Found in chert, silicified carbonate rocks, and petrified wood (see Table 3.6 for the gem varieties).
Potassic Feldspar (see Table 5.1)
Plagioclase Feldspar (see Table 5.1)
Biotite Mica (see Table 5.1)
Muscovite Mica (see Table 5.1)
Clay Minerals • Naturally white, light-gray, or green, but commonly stained by iron oxides in shades of brown, yellow, orange, or red. • White streak (unless stained by iron oxides). • Well-developed cleavage in one direction, but not seen because it occurs on a microscopic scale. • Translucent to pseudo-opaque with a nonmetallic greasy, silky, or dull luster. • Friable with a Mohs hardness of 1-2.5. • Greasy feel and commonly has a argillaceous or musty odor when breathed upon. • Swells and/or softens in water, excessively so for the montmorillonite variety. • Fusibility is 7. • Specific gravity is 2.00 to mainly 2.60-2.90. • Found mainly in mudrocks but also as an impurity in most other sedimentary rocks (see Table 3.6 for the principal sub-varieties: chlorite, illite, kaolinite, and montmorillonite). With increasing amounts of iron oxide impurities, the clay-rich mudrocks become gradational with red, brown, and yellow ochers.
CARBONATE MINERALS
Calcite • Usually colorless or white when pure, but commonly stained by iron oxides in shades of brown, yellow, or red; occasionally pale-blue, pale-green, pale-purple, gray, or black. • White to pale-gray streak. • Transparent to pseudo-opaque with a nonmetallic vitreous or dull luster. • Prominent cleavage in three directions, yielding rhombohedral fragments. • Brittle with a Mohs hardness of 3. • Commonly fluoresces red, pink, or white in both short- and long-wavelength, and blue in short-wavelength UV lights; and commonly phosphoresces bluish-white in both short- and long-wavelength UV lights. • Effervesces strongly in cold dilute HCl acid. • Fusibility is 7. • Specific gravity is 2.71-2.94. • Found mainly in carbonate rocks, but commonly occurs as veins or cement in most other sedimentary rocks (see Table 3.6 for sub-varieties).

Dolomite

- Usually colorless, white, gray, or pale-pink, but commonly with brownish or yellowish tinges due to common iron impurities.
- White streak.
- Transparent to translucent with a nonmetallic vitreous to pearly or dull luster.
- Prominent cleavage in three directions, yielding rhombohedral fragments.
- Brittle with a Mohs hardness of 3.5-4.
- Effervesces weakly in cold dilute HCl but strongly when powdered.
- Fusibility is 7.
- Specific gravity is 2.85-2.93.
- Found mainly in carbonate rocks, and only rarely as a cement in siliciclastic sedimentary rocks.

SULFATE AND HALIDE MINERALS

Anhydrite

- Usually colorless to light- or dark gray, occasionally pale-blue or pale-violet, but can be stained by iron oxides.
- White streak.
- Transparent to translucent with a nonmetallic vitreous to pearly luster.
- Prominent cleavage in one direction, but breaks mainly by uneven or splintery fracture.
- Brittle with a Mohs hardness of 3-3.5.
- Unlike gypsum, which it closely resembles, does not decrepitate in a candle flame, yield water condensate when heated in a test tube, or dissolve in hot dilute HCl acid.
- Fusibility is 3-3.5.
- Specific gravity is 2.89-3.00.
- Found only in evaporite rocks.

Gypsum

- Usually colorless, white, or light-gray, but can be stained by iron oxides.
- White streak.
- Transparent to translucent with a nonmetallic vitreous to pearly or silky luster.
- Prominent cleavage in one direction, but sometimes breaks by conchoidal or splintery fracture.
- Sectile and flexible but not elastic with a Mohs hardness of 2.
- Decrepitates in a candle flame, quickly yields water condensate when heated in a test tube due to dehydration to anhydrite, and readily dissolves in hot dilute HCl acid.
- Fusibility is 3-3.5.
- Specific gravity is 2.30-2.37.
- Found mainly in evaporite rocks, but can occur as veins or cement in other sedimentary rocks (see Table 3.6 for sub-varieties).

Halite

- Usually colorless or white, but when impure can be gray, yellow, red, blue, or purple.
- White streak.
- Transparent to translucent with a nonmetallic vitreous to greasy luster.
- Prominent cleavage in three directions, yielding cubical fragments.
- Brittle with a Mohs hardness of 2-2.5.
- Highly soluble in warm water and has a salty taste.
- Fusibility is 1.5.
- Specific gravity is 2.16-2.20.
- Found only in evaporite rocks or evaporitic encrustations.

OXIDE MINERALS

Hematite

- Red to brownish-red but occasionally purple or black.
- Red to reddish-brown streak.
- Pseudo-opaque to opaque with a submetallic to nonmetallic dull or silky luster.
- No cleavage, but breaks by uneven fracture.
- Brittle to friable or pulverulent with a Mohs hardness of 5-6.5.
- Weakly to moderately attracted to a neodymium magnet.
- Fusibility is 7.
- Specific gravity is 4.90-5.26.
- Principal mineral in ironstones, but commonly also found as a cement or stain in most other sedimentary rocks. Called red ocher when occurring in massive aggregates, commonly with clay mineral impurities.

Goethite

- Dark-brown, reddish- to yellowish-brown, brownish-yellow, or brownish-black.
- Brownish- or orangish-yellow to yellowish-brown streak.
- Pseudo-opaque to opaque with a submetallic to nonmetallic dull or silky luster.
- Prominent cleavage in one direction, but is usually not seen because it occurs on a microscopic scale.
- Brittle to pulverulent with a Mohs hardness of 5-5.5.
- Weakly to moderately attracted to a neodymium magnet, and when heated in a test tube yields water condensate due to dehydration to red hematite.
- Fusibility is 5.5.
- Specific gravity is 3.30-4.37.
- Commonly found as a cement or stain in most sedimentary rocks. Compositionally similar to limonite, a mineraloid (see below). Called brown ocher when occurring in massive aggregates, commonly with clay mineral impurities.

PHOSPHATE MINERALS

Apatite

- Usually greenish-yellow to bluish-, grayish- or yellowish-green; occasionally pale-pink, orange, brown, brownish-red, or red.
- White streak.
- Transparent to pseudo-opaque with a nonmetallic vitreous or greasy to mainly resinous luster.
- No prominent cleavage, but breaks by uneven or conchoidal fracture.
- Brittle with a Mohs hardness of 5.
- Commonly fluoresces yellow, brownish-yellow or greenish-yellow in both short- and long-wavelength UV lights.
- Occurs in finely granular aggregates that are commonly referred to as 'collophane' or 'fluorapatite'.
- Fusibility is 5.
- Specific gravity is 3.15-3.35.
- Found only in phosphatic sedimentary rocks.

MINERALOIDS

Opal

- Usually colorless or white; sometimes pale shades of gray, brown, red, or multi-colored.
- White streak.
- Transparent to mainly translucent with a nonmetallic vitreous to waxy or resinous luster.
- No cleavage, but breaks by conchoidal fracture.
- Brittle with a Mohs hardness of 5-6.
- Commonly fluoresces yellowish-green in both short- and long-wavelength, and pale-yellow in long-wavelength UV lights; and commonly phosphoresces white in both short- and long-wavelength UV lights.
- Unlike the other forms of silica (crystalline quartz and chalcedony), opal is amorphous and so is isotropic.
- Fusibility is 7.
- Specific gravity is 2.00-2.50.
- Found in petrified wood, siliceous sinter, and rarely chert (see Table 3.6 for the gem varieties).

Limonite
• Usually yellow to yellowish-brown; sometimes other shades of brown. • Yellow to yellowish-brown streak. • Translucent to opaque with a submetallic to nonmetallic dull or silky luster. • No cleavage, but breaks by uneven fracture. • Brittle to friable or pulverulent with a Mohs hardness of 4-5.5. • Weakly to moderately attracted to a neodymium magnet, and yields water condensate when heated in test tube due to dehydration to red hematite. • Fusibility is 5-5.5. • Specific gravity is 2.70-4.30. • Commonly found as a cement or stain in most sedimentary rocks. Called brown or yellow ocher when occurring in massive aggregates, commonly with clay mineral impurities.
Bitumen
• Black whole and streak colors. • Opaque with a nonmetallic dull or, occasionally, vitreous luster. • No cleavage, but breaks by uneven to conchoidal fracture. • Brittle with a Mohs hardness of < 2. • Quickly melts or burns in a candle flame, giving off dark fumes with a pungent petroliferous odor. • Fusibility is 0. • Specific gravity is about 1. • Found mainly in bituminous rocks, but occasionally occurs as impregnations in siliciclastic and carbonate sedimentary rocks.

Table 5.12: Simplified classification of silicic sedimentary rocks.

General Texture	Specific Texture	Composition[1]	Rock Variety[2]		Comments	
SILICICLASTIC ROCKS						
GRAVELSTONES						
clastic	grains mostly larger than 2.00 mm (gravel size)	gravel clasts are rock fragments of potentially any type	grain-supported	CONGLOMERATE	• Gravel clasts are mainly rounded.	• Pore spaces are filled with either cement (the same ones in sandstones, see below) or, more commonly, with a finer-grained sediment matrix (i.e. sand, silt, and/or clay). • Any color or combination of colors is possible depending on the rock fragments present, and the types of cement or matrix. • Adjectives can be added to rock names to indicate predominant clast type or size: e.g. pebble conglomerate and chert breccia. Such labeling is especially necessary to distinguish the igneous and sedimentary varieties of breccia. • A breccia or conglomerate with a sandy matrix is gradational with a pebbly sandstone.
				breccia	• Gravel clasts are mainly angular.	
			matrix-supported	diamictite	• Gravel clasts occur within a finer-grained sediment matrix composed of sand, silt, and/or clay, and can be of any shape or color. • Diamictites are gradational with pebbly sandstones and pebbly mudrocks.	

[1] Percentages are for the whole rock.
[2] The relative global abundance of rock varieties is indicated by the rock name font: **'ABUNDANT'**, 'COMMON', and 'rare'.

SANDSTONES						
clastic	grains mostly between 2.00 and 0.0625 mm (sand size)	<15% matrix	>90% of sand grains are quartz	**QUARTZ SANDSTONE** (a.k.a. quartz arenite)	• Rock is usually white or light-gray in color.	• Overall rock colors are largely determined by the dominant grain type and cement. However, all varieties can be colored in shades of brown, yellow, orange, red, or purple when iron oxides are present as either a cement or stain. Most sandstones are yellowish- to reddish-brown due to such staining. • Other grain types are occasionally abundant, such as flakes of muscovite mica (in **micaceous sandstone**), green glauconite (in **glauconitic sandstone**), and glass shards and pumice (in **volcanoclastic sandstone** = pyroclastic tuff). • Interstitial cements [and their descriptive adjectives] include: calcite [calcareous] and quartz [silicified] > iron oxides – hematite and less commonly goethite or limonite [ferruginous] and clay minerals [argillaceous] > dolomite [dolomitic], gypsum [gypsiferous], and siderite [sideritic]. The adjective is often added to the rock name: e.g. calcareous arkosic sandstone. Only quartz cement is common in quartz sandstones, and calcite cement predominates in the other varieties. • Most sandstones are porous to some degree and so absorb water when the surface is wetted. • Sandstones are usually friable with a Mohs hardness of less than 5.5, and are only harder than this when they have abundant quartz cement. The latter rocks are commonly referred to as **orthoquartzite**. • The specific gravity of sandstones (broadly defined) ranges from 2.17 to 2.70 g/cc and averages 2.40 g/cc with the wide spread in values reflecting the amount of porosity (i.e. the higher the porosity, the lower the specific gravity).
			<90% of sand grains are quartz with feldspar > rock fragments in the other grains	ARKOSIC SANDSTONE (a.k.a. arkose and feldspathic arenite)	• Rock is usually pinkish or reddish in color due to potassic feldspar, but occasionally is light-gray due to plagioclase. • Gradational with graywacke sandstone.	
			<90% of sand grains are quartz with rock fragments > feldspar in the other grains	LITHIC SANDSTONE (a.k.a. litharenite)	• Rock usually has a 'salt-and-pepper' coloration due to quartz and darker rock fragments, but can also be light- to dark-gray. • Gradational with graywacke sandstone.	
		>15% intergranular silt- and clay-rich matrix; sand grains can be of any composition but quartz predominates with feldspar and/or rock fragments usually common	graywacke sandstone (a.k.a. graywacke or greywacke)		• Overall colors are usually dark-gray or greenish-gray due to abundant clay minerals and secondary chlorite in the matrix. • As indicated by the presence of chlorite, this is often a metasedimentary rock (see Table 5.17). • Plagioclase is usually intermediate to calcic in composition and rock fragments are commonly volcanic. • Gradational with a sandy mudrock. • Commonly gives off an argillaceous or musty odor when breathed upon due to the high clay-mineral content.	

<table>
<tr><th></th><th colspan="5">MUDROCKS</th></tr>
<tr><td rowspan="3">clastic</td><td rowspan="3">grains mostly less than 0.0625 mm

(silt and clay size)</td><td>mainly quartz silt</td><td rowspan="2">non-fissile</td><td>SILTSTONE</td><td>• Resembles the various sandstone varieties in all respects except for grain size (see above for colors and cements).
• In comparison to the finer-grained mudrocks, siltstones tend to be harder (but still less than 5.5) and break into irregular or slab-like fragments rather than thin flakes as in shale.</td></tr>
<tr><td rowspan="2">a variable mix of silt-size quartz and minor feldspar, plus clay-size clay minerals with sand-size mica flakes common</td><td>MUDSTONE</td><td rowspan="2">• Overall colors are usually medium to dark shades of gray or green, but can be yellowish-, orangish- or reddish-brown, red, or purple depending on the nature of the iron-oxide stain, or black if carbonaceous. Mottled or spotty combinations of colors are common (as in the so-called variegated shales). Colors are lighter when the rock is silt-rich and darker when clay-rich.
• These rocks commonly exhibit finely laminated stratification.
• Both rock types are primarily lithified by compaction but can have minor amounts of cement, usually either calcite or iron oxides.
• Calcareous shale is gradational with marlstone (see Table 5.13).
• Can contain significant amounts of sand grains or even gravel clasts: e.g. pebbly mudstone and sandy shale.
• Will often emit an argillaceous odor (a musty smell like wet modelling clay) when breathed upon due to the rapid uptake of moisture from your breath. For the same reason, these rocks tend to stick to a dry tongue.
• Depending on the amount and type of clay mineral present, these rocks may have a smooth, greasy feel between the fingers.
• The ‘chew test’ can indicate the relative amounts of silt- and clay-size grains.
• Almost all these rocks (after powdering) yield water condensate when heated in a test tube due to the presence of clay minerals.
• Rocks rich in the clay mineral montmorillonite are called bentonite and will swell up to several times their dry volume when placed in water.
• The specific gravity of clay-rich mudrocks ranges from 2.06 to 2.66 g/cc and averages 2.25 g/cc.</td></tr>
<tr><td>fissile</td><td>SHALE</td></tr>
</table>

OTHER SILICIC ROCKS				
clastic	grains mostly less than 0.0625 mm (silt and clay size)	mainly opaline silica shells of marine or lacustrine diatoms	diatomite	• Typically white or light-gray in color. • Diatoms in diatomites have flat, circular forms, radiolarians in radiolarite have complex, net-like forms, and sponge spicules in spiculite have needle-like or other spikey forms. • Usually very soft and pulverulent, but when silicified can be harder than Mohs 5.5 whereupon it becomes a variety of chert (see below).
		mainly opaline silica shells of marine radiolarians	radiolarite	
		mainly opaline silica spicules from marine sponges	spiculite	
crystalline	aphanitic	mainly microcrystalline quartz and/or chalcedony	CHERT	• Usually the result of the recrystallization (a.k.a. silicification) of limestone and so can contain relict structures, including fossils and clastic grains. Incompletely silicified limestones are called **porcellanite**. • Almost any color is possible but light to dark shades of gray and pale-brown are most typical, and color banding or mottling is common. Dark-gray to black chert that occurs as nodules within limestone is called **flint**. When exhibiting vivid green, yellow, or red colors, or a combination of these or other colors, the rock is the gemstone **jasper**. See also the **banded iron formation** variety of ironstones in Table 5.13. • Mohs hardness is always greater than 5.5 and the rock has a conspicuous conchoidal fracture.
		opaline silica	siliceous sinter	• The result of direct precipitation in hot springs and geysers where it occurs as highly porous to dense encrustations. • Usually white in color unless tinted by various impurities. • Mohs hardness is highly variable given that the sinter can vary from a semi-cohesive sediment to a hard rock.

Table 5.13: Simplified classification of non-silicic sedimentary rocks.

<table>
<tr><th>General Texture</th><th colspan="3">Specific Texture and Composition</th><th>Rock Variety[1]</th><th>Comments</th></tr>
<tr><td colspan="6">CARBONATE ROCKS</td></tr>
<tr><td rowspan="6">clastic</td><td colspan="3">silt-size and finer grains of mainly calcite and clay minerals</td><td>marlstone
(a.k.a. marl)</td><td>• Gradational with calcareous shale.
• The specific gravity of marlstones ranges from 2.26 to 2.63 g/cc and averages 2.49 g/cc.</td></tr>
<tr><td rowspan="6">mainly calcite</td><td rowspan="3">abundant, conspicuous carbonate skeletal fragments (bioclasts) of fossilized marine or lacustrine invertebrates</td><td>sand- to gravel-sized fragments of any type of invertebrate</td><td>FOSSILIFEROUS
(or BIOCLASTIC)
LIMESTONE</td><td rowspan="6">• Almost all limestones are fossiliferous to some degree although this is not always obvious. If a specimen is not the coquina or chalk variety, then it can be referred to more broadly as fossiliferous or bioclastic limestone.
• Most limestones are white or light- to medium-gray, commonly with a yellowish or pinkish tint, but other colors are possible, depending on the impurities, including black and shades of green, blue, and brown.
• Many invertebrate fossils were originally composed of aragonite but, during diagenesis, this mineral is almost always replaced by calcite.
• Because calcite is highly soluble in groundwater, these rocks commonly have macroscopic pores (called ‘vugs’).
• Pore spaces between bioclasts are filled with calcite cement and/or fine-grained carbonate sediment: i.e. mud matrix (called ‘micrite’). The matrix has a dull luster and is pseudo-opaque with a dark-gray color.
• All carbonate rocks are softer than Mohs 5.5. Chalk is usually soft enough to be cut with a fingernail and is commonly pulverulent. Among the calcitic carbonates, those with crystalline textures are harder than those with clastic textures. Dolomitic carbonates are as hard, or harder, than the crystalline calcitic carbonates.
• All clastic calcitic carbonate rocks are highly susceptible to recrystallization with replacement of calcite by more calcite – making ‘recrystallized limestone,’, by dolomite – making ‘dolostone’, or by silica (quartz or chalcedony) – making ‘siliceous limestone’ or, eventually, ‘chert’.
• These rocks commonly exhibit ‘stylolites’, irregular, zig-zagging seams of clay minerals and other insoluble material.
• The specific gravity of limestones (broadly defined) ranges from 2.23 to 2.72 g/cc and averages 2.64 g/cc.</td></tr>
<tr><td>gravel-sized shells of molluscs</td><td>coquinoid limestone
(a.k.a. coquina)</td></tr>
<tr><td>sand- to mainly silt-sized shells of plankton (mostly foraminifera)</td><td>CHALK</td></tr>
<tr><td colspan="2">mainly silt-sized and finer grains of carbonate sediment</td><td>MICRITIC LIMESTONE
(a.k.a. lithographic limestone)</td></tr>
<tr><td colspan="2">mainly sand-size spherical to ovoid ooliths that have an internal concentric lamination</td><td>oolitic limestone
(a.k.a. oolite)</td></tr>
<tr><td>organic framework</td><td colspan="2">mainly an intergrowth of stationary colonial marine invertebrates</td><td>REEF LIMESTONE</td></tr>
</table>

[1] The relative global abundance of rock varieties is indicated by the rock name font: ‘**ABUNDANT**’, ‘COMMON’, and ‘rare’.

crystalline	phaneritic to mainly aphanitic calcite	**CRYSTALLINE LIMESTONE**		• From secondary recrystallization of limestone with partial or complete destruction of the original clastic or organic framework texture.
		travertine (a.k.a. calcite alabaster)	calcareous sinter	• From primary precipitation in subsurface caves and fissures. • Translucent with mainly white, or light-, yellowish- or reddish-brown colors, and with color-banding almost always present. • Dense (non-porous).
			calcareous tufa	• From primary precipitation in hot or cold springs, or waterfalls. • White or light-gray to light-yellow or light-brown in color. • Highly porous with a spongy appearance and commonly well layered.
	phaneritic to mainly aphanitic dolomite	DOLOSTONE (a.k.a. dolomite)		• Almost always from the recrystallization (a.k.a. dolomitization) of limestone. • White to mainly light-pinkish- to brownish-gray in color. • Calcite effervesces vigorously when touched by dilute hydrochloric acid whereas dolomite does not react unless first powdered. Intermediate reactions due to mixtures of the two minerals will allow recognition of limestone, dolomitic limestone, calcitic dolostone, or dolostone (see section 3.2.8 for more information). • Alizarin Red S dye will stain calcite either pink or red, but will leave dolomite unaffected (see section 3.2.9 for details). • The specific gravity of dolostones ranges from 2.77 to 2.80 g/cc and averages 2.79 g/cc.

<table>
<tr><th colspan="5">EVAPORITE ROCKS</th></tr>
<tr><td rowspan="3">crystalline</td><td>phaneritic to mainly aphanitic halite</td><td colspan="2">rock salt</td><td>• Colorless or white to yellowish-white in color.
• Has a salty taste.
• Readily dissolves in warm water (especially if first powdered).
• Isotropic in thin slivers.
• Decrepitates in a candle flame.
• Colors a blowpipe flame deep yellow.
• Mohs hardness of the pure, massive rock salt is 2-2.5.
• The specific gravity of pure, massive rock salt ranges from 2.16 to 2.20 g/cc.</td></tr>
<tr><td>phaneritic to mainly aphanitic anhydrite</td><td>rock anhydrite (a.k.a. anhyrock)</td><td rowspan="2">The 'rock' prefix is optional but should be used when there is the possibility of confusion with the minerals anhydrite and gypsum</td><td>• White or light- to dark-gray in color.
• Easily fuses in a blowpipe flame to an opaque white bead.
• Colors a blowpipe flame brownish-red.
• Mohs hardness of pure, massive rock anhydrite is 3-3.5.
• The specific gravity of pure, massive rock anhydrite ranges from 2.89 to 3.00 g/cc.
• Gradational with rock gypsum.</td></tr>
<tr><td>phaneritic to mainly aphanitic gypsum</td><td>ROCK GYPSUM (a.k.a. gyprock)</td><td>• White or light-gray in color.
• Includes the selenite, satin-spar, and alabaster varieties.
• Readily dissolves in hot dilute HCl acid.
• Decrepitates in a candle flame.
• Quickly yields water condensate in a heated test tube due to dehydration to anhydrite.
• Colors a blowpipe flame yellowish-red.
• Commonly fluoresces white to bluish-white (short) or white to yellowish-white (long) in ultraviolet lights.
• Mohs hardness of pure, massive rock gypsum is 2.
• The specific gravity of pure, massive rock gypsum ranges from 2.30 to 2.37 g/cc.
• Gradational with rock anhydrite.</td></tr>
<tr><td colspan="5">Other evaporate rocks with different mineralogies exist but are exceedingly rare.</td></tr>
</table>

CARBONACEOUS ROCKS					
amorphous	semi-carbonized woody plant remains		lignite (a.k.a. brown coal)	• Brown to brownish-black in color, and soft with friable surfaces, a hackly fracture, and dull luster. • Gem-quality lignite is known as **jet** and, relative to the normal rock, is much tougher (Mohs hardness of up to 4 with greater impact resistance), and has a well-developed conchoidal fracture and waxy luster.	• From accumulation and transformation of terrestrial plant remains into bitumen. Commonly contains plant fossils, especially in lignite. • Both burn in a candle flame, coloring it yellow and emitting dark smoke and a petroliferous odor. • Mohs hardness varies from <1-2 for normal lignite and 2-3 for bituminous coal. • Specific gravity ranges from 1.0 to 1.4 g/cc for lignite, and 1.2 to 1.5 g/cc for bituminous coal.
	fully carbonized woody plant remains		bituminous coal	• Black in color, usually with alternating shiny and dull layers, and much harder than lignite with a hackly to conchoidal fracture. • Tends to break along parallel fractures that are nearly perpendicular to bedding. • Handling the rock leaves a black, powdery residue on the fingers.	
FERRUGINOUS ROCKS					
crystalline or clastic	coarse-grained (i.e. phaneritic or sand/gravel-size) to mainly fine-grained (i.e. aphanitic or silt-size)	abundant (>20%) iron oxides (hematite, goethite, limonite, or magnetite) and/or other iron-rich minerals	ironstone	• Usually from the infusion of iron oxides into any pre-existing porous sedimentary rock. • Red, reddish-brown to brownish-black in color. • Commonly rich in secondary silica and gradational with ferruginous chert. • A **banded iron formation** is a rare variety of ironstone consisting of alternating, thin bands of chert (gray or, if ferruginous, red) and either hematite (red) or magnetite (dark-gray or black). It is considered a low-grade metamorphic rock in some classifications. The ferruginous chert is a variety of the gemstone **jasper**. • Commonly occurs in stalactite-like formations in caves and other rock cavities. • The specific gravity of pure, massive iron oxide (goethite, hematite, or magnetite) ranges from 3.30 to 5.20 g/cc.	

PHOSPHATIC ROCKS				
crystalline or clastic	coarse-grained (i.e. phaneritic or sand/gravel-size) to mainly fine-grained (i.e. aphanitic or silt-size)	abundant (>20%) phosphatic minerals, including collophane, fluorapatite, and related minerals	phosphorite	• From recrystallization of marine or lacustrine sediments with abundant phosphate-rich fish bones and/or fossil feces (coprolites). • Usually dark-brown to brownish-black in color with a shiny luster. • The specific gravity is on the order of 3.1 g/cc, higher than all the other sedimentary rocks except ironstone.

Table 5.14: Identification key for common macroscopic invertebrate fossils in sedimentary rocks (modified after Matthews 1960: 24-31 with additions from other sources).

I. Fossil displays radial symmetry (i.e. a symmetrical repetition of parts around a central axis)	**Invertebrate animal[1]**
A. Fossil has a tapering, cylindrical, or cone shape.	
1. Has internal, longitudinal, radial partitions and is cone-shaped.	coral[6]
2. Has internal, transverse partitions and tapers at one end.	cephalopod[12]
3. Has no internal partitions.	
a. Shell is large and heavy, and usually has external longitudinal ribs.	rudistid[11]
b. Shell is small (usually less than 5 cm long), tusk-shaped, and open at both ends.	scaphopod[14]
B. Fossil has a disk-like or flattened dome shape.	
1. Has a radiating star pattern on top.	echinoid[9]
2. Has a subconical to hemispherical dome shape, a concave or flat base, and minute pits covering the surface, and is small (usually less than 8 cm across).	bryozoa[5]
3. Has a disk-like shape and is small (less than 1.5 cm).	foraminifer[16]
4. Has a disk- or button-like shape with internal longitudinal, radial partitions.	coral[6]
C. Fossil consists of many segments or plates.	
1. Has circular disks or chambers which, when united, form a cylinder.	
a. Has a tapered shell.	cephalopod[12]
b. Has small, disk-shaped segments of similar thickness with a hold through the middle.	crinoid[8]
2. Has many-sided plates.	
a. Bud-shaped with 13 wedge-shaped plates.	blastoid[7]
b. Cup-shaped with many curved plates surrounded by branching arms.	crinoid[8]
II. Fossil displays bilateral symmetry (i.e. a symmetrical duplication of parts on each side of a bisecting plane)	
A. Fossil is coiled within a single plane.	
1. Shell is divided by internal, suture-like, transverse partitions.	cephalopod[12]
2. Shell has no internal partitions.	gastropod[13]
3. Shell is tiny and spindle-shaped (resembles a wheat grain in form and size).	foraminifer[16]
B. Fossil is not coiled.	
1. Shells are two valves attached along a hinge.	
a. Plane of symmetry is parallel to the hinge line (both valves are nearly identical).	pelecypod[11]

b. Plane of symmetry is perpendicular to the hinge line (the two valves are not identical), and the shell exterior is strongly radially ribbed and has 'ears' at each end of the hinge.	pelecypod[11]
c. Plane of symmetry is perpendicular to the hinge line (the two valves are not identical), and the shell has no 'ears' but the larger of the two valves commonly has an opening and a 'beak-like' projection at the hinge.	brachiopod[4]
2. Fossil has a tapering, cylindrical, or cone shape.	
a. Has internal, longitudinal, radial partitions, and is cone-shaped.	coral[6]
b. Has internal, transverse partitions and tapers at one end.	cephalopod[12]
c. Has no internal partitions.	
1. Shells are large and heavy, and usually have external, longitudinal ribs.	rudistid[11]
2. Shell is small (usually less than 5 cm long), tusk-shaped, and open at both ends.	scaphopod[14]
3. Fossil is heart-shaped, domed, or flattened with a radial star pattern on top.	echinoid[9]
4. Fossil is segmented.	
a. It is divided into three longitudinal lobes, has a broad head and may be curled up.	trilobite[3]
b. It is flattened or elongate, and resembles modern-day shrimp, crab, or crayfish.	crustacean[2]
III. Fossil displays no apparent symmetry	
A. Fossil has no transverse internal partitions.	
1. Shell consists of two valves (but only the larger one is usually found) and is coiled like a ram's horn; it is low-spired, and has a very large opening and concentric ridges on the exterior.	pelecypod[11]
2. Shell consists of one valve and is tightly coiled; and in comparison with '1' just above, is higher-spired with a smaller opening and a smoother exterior.	gastropod[13]
B. Fossil is coiled but the coiling is not in one plane.	
1. Shell has transverse internal partitions.	
a. Shell is thick-walled with smooth, internal partitions and an irregular, loose coil; usually occurs in large compact masses of many individual shells.	rudistid[11]
b. Shell is thin-walled with suture-like internal partitions that are usually wrinkled and has a regular, tight coil; usually occurs as well-separated, individual shells.	cephalopod[12]
2. Shell has no transverse internal partitions.	gastropod[13]
3. Shell has a solid, spiral ridge around its central axis that resembles a corkscrew.	bryozoa[5]
C. Fossil has no coil.	

1. Fossil resembles a narrow saw blade and is usually preserved as a thin carbon film.	graptolite[10]
2. Fossil has an irregular cone-shape with internal, longitudinal, radial partitions.	coral[6]
3. Fossil shell has two asymmetrical valves and resembles a clam or oyster.	pelecypod[11]
4. Fossil has twig-like branches.	
a. Its surface is densely covered with minute openings.	bryozoa[5]
b. It has relatively large, evenly distributed openings with internal, longitudinal, radial partitions.	coral[6]
5. Fossil is lace-like and occurs as thin sheets or films.	bryozoa[5]
6. Fossil is composed of radiating masses of polygonal or circular tubes containing internal radial partitions.	coral[6]
7. Fossil is irregular in form but is roughly cylindrical or conical with a rough surface.	
a. Has a thick wall and large axial opening with usually longitudinal, external ribs.	rudistid[11]
b. Has no large axial opening but has a surface densely covered by small pits with internal, needle-like spicules which are the only hard parts (impressions of the soft parts are rarely preserved).	sponge[15]

[1] With one exception, all fossils listed here have internal or external skeletons composed of calcium carbonate (CaCO3), with this always calcite except for some geologically recent fossils which are composed of aragonite. The exception is sponge, the spicules of which are almost always composed silica (SiO_2), either microcrystalline quartz or chalcedony, or opal in geologically recent fossils.
[2] Phylum Arthopoda, superclass Crustacea.
[3] Phylum Arthopoda, class Trilobita.
[4] Phylum Brachiopoda, classes Articulata and Inarticulata.
[5] Phylum Bryozoa, class Stenolaemata.
[6] Phylum Cnidaria, class Anthozoa (a.k.a. coral).
[7] Phylum Echinodermata, class Blastoidea.
[8] Phylum Echinodermata, class Crinoidea (a.k.a. sea lilies).
[9] Phylum Echinodermata, class Echinoidea (a.k.a. sea urchins and sand dollars).
[10] Phylum Hemichordata, class Graptolithina.
[11] Phylum Mollusca, class Bivalvia (a.k.a. clams, oysters and mussels).
[12] Phylum Mollusca, class Cephalopoda (a.k.a. ammonoids and nautiloids).
[13] Phylum Mollusca, class Gastropoda (a.k.a. snails).
[14] Phylum Mollusca, class Scaphopoda (a.k.a. tusk shells).
[15] Phylum Porifera, numerous classes.
[16] Not an animal, but rather a single-celled, amoeboid protist.

5.4 Metamorphic Rocks

5.4.1 Description

The most challenging aspect of metamorphic rocks is their highly varied mineralogy. This reflects not only the composition of the original parent rock, but also the elements brought in by mineralizing solutions, and the pressure and temperature conditions during metamorphism. The metamorphic process involves the recrystallization of minerals already in the parent rock and, usually, the crystallization of new minerals by a recombination of elements in the pre-existing minerals and/or the addition of new elements by invading solutions. The end result is the creation of many new minerals that do not occur in igneous rocks. Some of these will survive the weathering and erosional processes to be incorporated within siliciclastic sedimentary rocks, but only as microscopic grains, except in the case of gravel-size clasts of metamorphic rocks in breccias and conglomerates. Table 5.15 lists the most common metamorphic minerals along with tips for their macroscopic identification. One or more of these minerals will almost always be present in a metamorphic rock. In the rare instances when they are absent, it is because the parent rock was monomineralic and no outside elements were introduced (e.g. pure limestone transformed into calcitic marble). There are still more varieties of metamorphic minerals but these are only rarely seen. Two of them, however, are archaeologically significant: jadeite (in 'jade'), and lazurite/haüyne (in 'lapis lazuli'; see Tables 3.6 and 3.5, respectively). In addition to the metamorphic minerals and, in fact, dominating the mineral assemblages are the by-now familiar varieties found in igneous and sedimentary rocks: felsic minerals – quartz, and potassic and plagioclase feldspars; mafic minerals – biotite and muscovite micas, hematite, hornblende amphibole, magnetite, marcasite, olivine, and pyrite; and carbonate minerals – calcite and dolomite (Table 5.16). Feldspathoid minerals will not be present as these do not survive when a feldspathoid-rich parent rock is metamorphosed.

5.4.2 Classification

Metamorphic rocks are too compositionally and structurally diverse to be amenable to a straight-forward combined textural and mineralogical classification, such as the one for igneous rocks. Unlike for igneous rocks, however, there is little disagreement among geologists on the general classification of metamorphic rocks. Most schemes use the same root names, which are based primarily on fabric (a combination of grain texture and three-dimensional internal structure) and/or mineralogy. The classification of metamorphic rocks from the International Union of Geological Sciences (IUGS), which is employed in this book, is actually two independent but compatible schemes – one 'systematic' and the other 'specific' – with both suitable for macroscopic use (Fettes and Desmons 2007). The systematic scheme divides metamorphic rocks into the following three groups based on their internal structure: 'schist', 'gneiss', and 'granofels'. Rocks in the schist group exhibit well-developed 'schistosity' or 'schistose structure': i.e. elongated and/or platy grains are abundant and show a high degree of preferred orientation, which causes the rock to split parallel to this orientation into chips and slabs less than 1 cm thick (Figures 4.5d and 5.9-5.10). Rocks in the gneiss group have either no or only poorly developed schistosity, but exhibit a 'gneissose structure': i.e. grains are segregated into alternating, parallel bands or streaks of light (mainly felsic) and dark (mainly mafic) minerals, and the rock breaks either irregularly or into slabs more than 1 cm thick (Figures 4.5c and 5.11). All metamorphic rocks lacking either a schistose or gneissose structure belong to the granofels group (Figures 5.12-5.13). These groups were also recognized in the older classifications, but were divided into 'foliated' (schist and gneiss groups) and 'nonfoliated' (granofels group) rocks. Among the foliated rocks there is a strong interrelationship between type of foliation, mineralogy, and metamorphic grade (i.e. the temperatures and pressures of formation; see Figure 4.4).

The IUGS's systematic classification of a metamorphic rock uses one of the three structural groups as a root name and then adds, as a prefix, the predominant minerals in a hyphenated string from least to most abundant (left to right): e.g. garnet-biotite-quartz gneiss. There are two problems with this approach, however. First, it requires the recognition of the often difficult to identify metamorphic minerals. And second, it produces rock names that are not well accepted by most geologists: e.g. 'calcite granofels' in place of the traditional 'marble'.

For those uncomfortable with the systematic classification, the IUGS recommends its specific scheme which utilizes many familiar rock names from older classifications, but with some slight redefinitions. These names are based on grain texture, internal rock structure, mineralogy, and/or the type of parent rock, but these parameters are not applied in a systematic manner. As a consequence, some rock names are based only on texture (e.g. cataclastite), structure (e.g. schist), mineralogy (e.g. marble), parent rock (e.g. metaconglomerate), or a combination of these characteristics (e.g. quartzite). Although the specific classification may seem the more complicated of the two IUGS schemes, it is the easiest to use for non-geologists, as well as for many geologists. It is, therefore, the classification of metamorphic rocks used in this book as described in Table 5.17. This table also includes the group/root names from the systematic classification for its different sections. The specific rock names in this table will usually be the only ones applied to archaeological stones, but if the predominant mineralogy can be recognized, then it is useful to add this to the rock name: e.g. chlorite schist, or garnet-bearing (or garnetiferous) calcite marble.

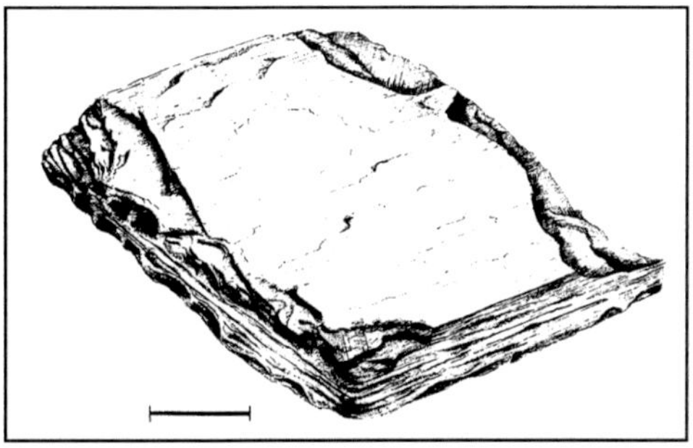

Figure 5.9: Example of slate with the foliation shown by a tendency to break into thin sheets and slabs. From Figure 42A in Raymond (1984) and with permission of Waveland Press, Inc. Scale bar = 2 cm.

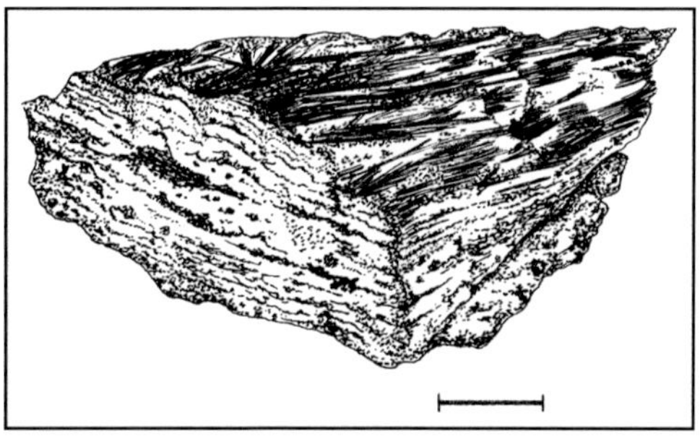

Figure 5.10: Example of schist with the foliation shown by the nearly parallel orientation of platy and acicular grains. From Figure 6 in Raymond (1984) with permission of Waveland Press, Inc. Scale bar = 2 cm.

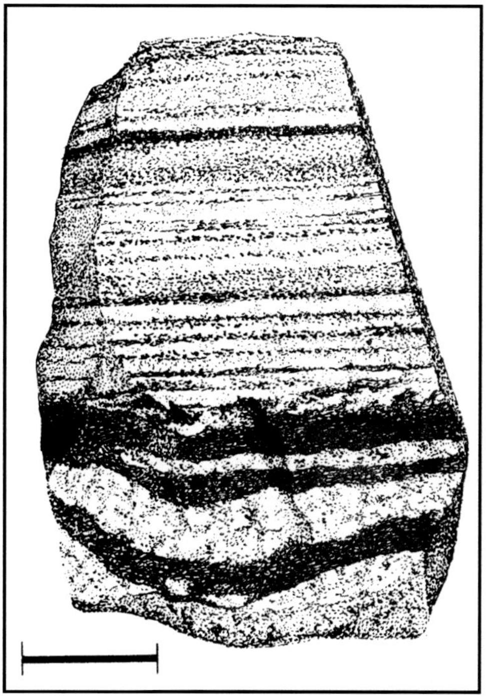

Figure 5.11: Example of gneiss with the foliation shown by the alternating layers of light- and dark-colored minerals. From Figure 42C in Raymond (1984) with permission of Waveland Press, Inc. Scale bar = 2 cm.

There is great compositional and textural variation among the non-foliated rocks, with some of these given special names (e.g. marble; Figure 5.12) and others using the igneous or sedimentary name of their parent rock, but with the 'meta' prefix added (e.g. metadiorite; Figure 5.13). In practice, however, geologists often leave off this prefix when the metamorphic grade is low and, consequently, the macroscopic appearance of the rock is not significantly different from that of the original (unmetamorphosed) parent rock. The low-grade 'metasedimentary' rocks often exhibit, macroscopically, what appears to be the original clastic texture, but the partial transformation to a crystalline texture is evident at the microscopic scale. At the finer levels of classification, other textural and compositional characteristics are used to determine the names of the non-foliated rocks. If the metamorphosing temperature becomes too high, the parent rock will melt, either entirely or partially. In the former case, a new magma is produced, and in the latter a hybrid rock, called 'migmatite', is created, being part metamorphic and part plutonic igneous.

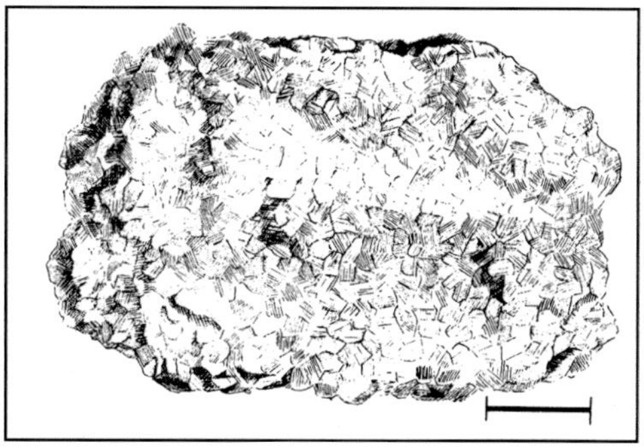

Figure 5.12: Example of calcitic marble. From Figure 43A in Raymond (1984) and with permission of Waveland Press, Inc. Scale bar = 2 cm.

Similar to porphyritic igneous rocks, there can be a conspicuous size dichotomy among the grains within some metamorphic rocks; the larger grains (often well-formed crystals) being called 'porphroblasts' rather than phenocrysts (Figure 5.14). These will usually be one of the metamorphic minerals in Table 5.15 but other, more common, minerals can also occur in this guise. In naming metamorphic rocks, one normally adds a prefix to the root name that includes the most abundant or conspicuous minerals in order of increasing abundance from left to right (e.g. quartz-muscovite schist). However, in the case of gneiss, it is also common to use the name of the mineralogically equivalent phaneritic igneous rock as a compositional descriptor (e.g. granite gneiss). The latter approach results not only in shorter names and easily recognizable compositions, but the named igneous rock is usually also the parent rock for the gneiss.

Figure 5.13: Example of metadiorite with quartz, plagioclase, and hornblende. From Figure 43B in Raymond (1984) with permission of Waveland Press, Inc. Scale bar = 2 cm.

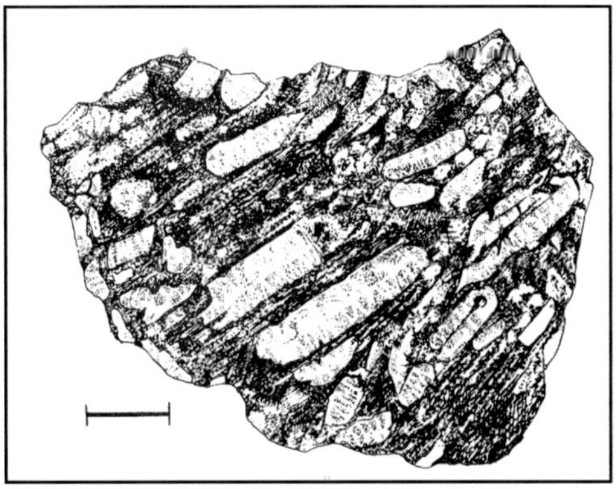

Figure 5.14: Example of porphroblastic schist with potassic feldspar porphyroblasts in a quartz-biotite matrix. From Figure 42D in Raymond (1984) with permission of Waveland Press, Inc. Scale bar = 2 cm.

5.4.3 Identification Key

The following key, when used together with Table 5.17, will allow you to identify the rock group and, sometimes, also a specific rock variety for an unknown metamorphic specimen.

1. Does the rock have a conspicuous phaneritic crystalline texture?
   If yes, go to 2.
   If no, go to 10.
2. Does the rock exhibit conspicuous compositional banding or streaking (i.e. alternating bands or streaks of light-colored felsic minerals and dark-colored mafic minerals)?
   If yes, go to 3.
   If no, go to 4.
3. Are there highly contorted, discontinuous bands of dark-colored mafic minerals that are surrounded by patches, veins, or contorted discontinuous bands of light-colored felsic minerals?
   If yes, the rock is migmatite.
   If no, the rock is either gneiss or amphibolite (see Table 5.17 to distinguish between these).
4. Does the rock exhibit a conspicuous directional structure (i.e. the elongated and/or platy grains have a parallel orientation)?
   If yes, the rock is either schist or phyllite (see Table 5.17 to distinguish between these).
   If no, go to 5.

5. Does the rock consist predominantly of calcite or dolomite?
   If yes, the rock is probably marble but also possibly skarn (see Table 5.17 to distinguish between these.
   If no, go to 6.
6. Does the rock consist of dark green grains (mainly minerals of the pyroxene group) and large, well-formed crystals of red garnet?
   If yes, the rock is eclogite.
   If no, go to 7.
7. Does the rock resemble one of the igneous phaneritic rock varieties, and lack compositional banding/streaking or a directional structure, but contain one or more metamorphic minerals?
   If yes, the rock is granulite or a phaneritic metaigneous rock.
   If no, go to 8.
8. Does the rock have a smooth, greasy feel and a Mohs hardness of less than 2.5?
   If yes, the rock is steatite.
   If no, go to 9.
9. Does the rock have a pseudo-clastic texture?
   If yes, go to 10.
   If no, and it has either an aphanitic crystalline or amorphous texture, go to 12.
10. Does the rock resemble a sedimentary breccia or conglomerate, but contain one or more metamorphic minerals?
    If yes, the rock is either metabreccia or metaconglomerate (see Table 5.17 to distinguish between these).
    If no, go to 11.
11. Does the rock resemble a sedimentary sandstone, but contain one or more metamorphic minerals and/or it breaks through the sand grains?
    If yes, the rock is metasandstone or, if it consists almost entirely of quartz, quartzite.
    If no, go to 12.
12. Are the grains within the rock fragmented or deformed?
    If yes, the rock is either cataclasite or mylonite (see Table 5.17 to distinguish between these).
    If no, go to 13.
13. Does the rock readily break into thin (< 1 cm thick) slabs?
    If yes, the rock is either slate or phyllite (see Table 5.17 to distinguish between these).
    If no, go to 14.
14. Does the rock have a grayish or blackish color?
    If yes with a light- to dark-gray or black color and dull luster, the rock is probably hornfels.
    If yes with a dark-gray or black color and a lustrous or iridescent sheen, the rock is anthracite coal.
    If no, go to 15.
15. Is the rock dark blue or dark greenish- to purplish-blue and has white veins or patches of calcite and/or minute golden specks of pyrite?
    If yes, the rock is lapis lazuli.
    If no, go to 16.
16. Is the Mohs hardness over 5.5?
    If yes, go to 17.
    If no, go to 19.

17. Does the rock consist predominantly of green minerals?
    If yes, the rock is either greenstone or jade (see Table 5.17 to distinguish between these).
    If no, go to 18.
18. Does the rock consist of white or a mottling of white and green minerals?
    If yes, the rock is jade.
    If no, the rock is probably hornfels.
19. Does the rock consist predominantly of calcite or dolomite?
    If yes, the rock is probably marble but possibly skarn (see Table 5.17 to distinguish between these).
    If no, go to 20.
20. Is the Mohs hardness of the rock less than 2.5 and does it feel greasy?
    If yes, the rock is steatite.
    If no, the rock is probably serpentinite.

5.5 'Stones' That Are Neither Rock Nor Mineral

The bits of 'stone' caught on archaeological screens include not only rocks and minerals, but also potentially many other solid materials. These can include fragments of objects made from plants (e.g. amber and bitumen), animals (e.g. mollusc and tortoise shells, ostrich eggshell, coral, pearl, antler, horn, bone, and ivory), or manufactured materials (e.g. glass, faience, ceramics, and metals). Although this book is only concerned with the identification of rocks and minerals, it is necessary to know how to distinguish these from other materials with which they sometimes occur. Table 5.18 provides some tips on how to do this for the most common non-geological materials.

Table 5.15: Common metamorphic rock-forming minerals and tips for their macroscopic identification.

Actinolite-Tremolite solid-solution series (amphibole group)

- Pale- to dark-green or brownish-green for actinolite, and colorless, white or light-gray for tremolite (but pink or pale-violet when Mn-bearing).
- White streak.
- Transparent to translucent with a nonmetallic vitreous or, when fibrous, silky luster.
- Prominent cleavage in two directions at ~56° and ~124° angles.
- Usually in acicular to long bladed prismatic crystals that are commonly in radiated groups; and occasionally in massive, granular to fibrous aggregates (called **nephrite** or, when gem quality, **jade**).
- Brittle, except for nephrite which is tough, with a Mohs hardness of 5-6.
- Actinolite is weakly attracted to a neodymium magnet, and tremolite is not attracted unless iron-bearing and then only weakly.
- Fusibility is 4.
- Specific gravity is 2.98-3.48.
- Found in some greenstones, skarns, schists, and marbles.

Andalusite

- Usually white, pink to red, or reddish- to purplish-brown; less commonly, green, yellow, or, when it has abundant carbonaceous impurities, dark-gray to nearly black.
- White streak.
- Transparent to mainly translucent or, when carbonaceous, opaque with a nonmetallic vitreous to dull luster.
- No prominent cleavage, but breaks by uneven to sub-conchoidal fracture.
- Brittle with a Mohs hardness of 6.5-7.5.
- Almost always in short prismatic crystals, commonly in radiated groups, with square cross-sections and flat ends; crystal cross-sections usually exhibit a black cross (due to carbonaceous impurities) within a white square.
- Fusibility is 7.
- Specific gravity is 3.14 to 3.20.
- Found in some hornfels.

Anthophyllite (amphibole group)

- Usually shades of brown, especially dark-yellowish-brown, and less commonly white, gray, or green.
- White streak.
- Transparent to mainly translucent or pseudo-opaque with a nonmetallic vitreous to pearly or, when fibrous, silky luster.
- Prominent cleavage in two directions at ~56° and ~124° angles.
- Brittle with a Mohs hardness of 5.5-6.
- Usually in parallel or radiated groups of acicular crystals or in massive, fibrous aggregates.
- Fusibility is 5-6.
- Specific gravity is 2.85 to 3.57.
- Found in some gneisses and schists.

Chlorite (clay and mica groups)

- Usually various shades of green; rarely brown, yellow, pink, or red.
- Transparent to translucent with a nonmetallic vitreous to pearly or dull luster.
- White to pale-green streak.
- Prominent cleavage in one direction with easy separation.
- Flexible but not elastic with a Mohs hardness of 2-3.
- Usually in massive, scaly aggregates; and rarely in pseudo-hexagonal tabular crystals.
- Greasy feel.
- Fusibility is 5-5.5.
- Specific gravity is 2.6-3.3.
- Found in some greenstones, phyllites, serpentinites, slates, steatites, and schists.

Chloritoid

- Usually dark-green or greenish-gray to greenish-black; occasionally bluish-gray.
- Translucent with a nonmetallic vitreous to pearly luster.
- White streak.
- Prominent cleavage in one direction.
- Brittle with a Mohs hardness of 6.5.
- Usually in massive, foliated aggregates with radial splays; rarely in tabular, pseudo-hexagonal crystals.
- Resembles chlorite, after which it is named, but is easily distinguished from it by its much greater hardness.
- Fusibility is 6-7.
- Specific gravity is 3.4 to 3.8.
- Found in some schists.

Cordierite

- Usually various shades of blue or violet (called **iolite** when gem quality); occasionally colorless, bluish-gray, yellow, or brown.
- Transparent to mainly translucent with a nonmetallic vitreous to resinous luster.
- White streak.
- No prominent cleavage, but breaks by sub-conchoidal fracture.
- Brittle with a Mohs hardness of 7-7.5.
- Commonly in short pseudo-hexagonal crystals; and also commonly in massive, granular aggregates.
- Fusibility is 5-5.5.
- Specific gravity is 2.53-2.78.
- Found in some gneisses, phyllites, schists, and slates.

Diopside (pyroxene group)

- Usually white or light-green; less commonly dark-green.
- Transparent to translucent with a nonmetallic vitreous luster.
- White to pale-gray streak.
- No prominent cleavage, but breaks by uneven fracture.
- Brittle with a Mohs hardness of 5.5-6.5.
- Usually in short, prismatic crystals with a square or octagonal cross-section; occasionally in massive to stalactitic granular aggregates.
- Fusibility is 4.
- Specific gravity is 3.22 to 3.25.
- Found in eclogite, and some marbles and skarns.

Epidote

- Yellowish-green, brownish-green, or blackish-green; rarely colorless, pale-yellow, or gray.
- Transparent to mainly translucent or pseudo-opaque with a nonmetallic vitreous luster.
- White to pale-gray streak.
- Prominent cleavage in one direction, but also breaks by uneven fracture.
- Brittle with a Mohs hardness of 6-7.
- Commonly in bladed to acicular prismatic crystals that are usually in radiated groups; and also commonly in massive, granular to fibrous aggregates.
- Sometimes weakly attracted to a neodymium magnet.
- Fusibility is 3-4.
- Specific gravity is 3.21-3.49.
- Found in some greenstones, marbles, schists, and skarns. Also a rare mineral in some plutonic igneous rocks, although some geologists would consider these to be metaigneous rocks.

Garnet Group with **almandine, andradite, grossular, hydrogrossular, pyrope, spessartine,** and **uvarovite**

- Most commonly almandine – red, purplish-red, purple, reddish-brown, or brownish-black; pyrope – red to purplish-red or nearly black; or grossular – greenish-white, pale- to dark-green, less often colorless, yellow, pink, reddish-orange, or brown.
- Occasionally spessartine – orange to reddish-orange, dark-red to brownish-red, brown, or black; andradite – yellowish-green, brownish-green, brown, greenish-brown, yellow, or black; uvarovite – dark- to vivid-green; or hydrogrossular – green to bluish-green, white, gray, or pink.
- White streak.
- Transparent to translucent (pyrope), translucent only (almandine, grossular, spessartine, and uvarovite), or pseudo-opaque to opaque (andradite and hydrogrossular) with a nonmetallic vitreous to resinous luster; transparent and semi-transparent crystals are isotropic.
- No cleavage, but breaks by conchoidal or uneven fracture.
- Brittle with a Mohs hardness of 7-7.5 (except 6-7 for hydrogrossular).
- Usually as dodecahedral or other multi-faceted, equidimensional (roughly spherical) crystals both within the groundmass and as porphroblasts.
- Iron-rich almandine and andradite are weakly to mainly moderately or strongly attracted to a neodymium magnet, and manganese-rich spessartine is only weakly attracted.
- Fusibility is 3-4, except uvarovite which is 7.
- Specific gravity is 3.58-4.32 g/cc.
- Found in eclogite, and some gneisses, marbles, schists, and skarns. Also a rare mineral in some plutonic igneous rocks, although some geologists would consider these to be metaigneous rocks.

Graphite

- Dark-gray to black.
- Opaque with a metallic to submetallic luster.
- Shiny black streak.
- Prominent cleavage in one direction with easy separation.
- Flexible and sectile with a Mohs hardness of 1-2.
- Usually in massive, scaly aggregates; and rarely as hexagonal tabular crystals in radiated groups.
- Has a greasy feel.
- Fusibility is 7.
- Specific gravity is 2.00 to 2.23.
- Found in some marbles, phyllites, schists, slates, and steatites.

Kyanite

- Usually mottled light- and dark-blue; occasionally white, gray, green, grayish-green, pink, yellow, or black.
- Transparent to pseudo-opaque with a nonmetallic vitreous to pearly luster.
- White streak.
- Prominent cleavage in one direction.
- Flexible with a variable Mohs hardness of 4-5.5 parallel to and 6-7 perpendicular to crystal length.
- Almost always in tabular to mainly long, bladed crystals that are commonly bent or twisted.
- Fusibility is 7.
- Specific gravity is 2.00-2.23.
- Found in some schists.

Serpentine Group with **antigorite**, **chrysotile**, and **lizardite**

- Usually white or various shades of green or yellow and commonly a mottling of these colors; occasionally gray, brown, or reddish-brown.
- Translucent to pseudo-opaque with a nonmetallic waxy or greasy luster when granular or, when fibrous, a silky luster.
- White streak.
- No prominent cleavage, but breaks by uneven or splintery fracture.
- Tough with a Mohs hardness of 2.5-5.
- Always in massive aggregates that are either granular (the antigorite and lizardite varieties) or fibrous (the asbestiform chrysotile variety).
- Commonly fluoresces yellow or white in both short- and long-wavelength UV lights.
- Fusibility is 7.
- Specific gravity is 2.55-2.61.
- Found in serpentinite, and some steatites and marbles. Also a rare mineral in some plutonic igneous rocks, although some geologists would consider these to be metaigneous rocks.

Sillimanite

- Usually colorless, white to light-gray, or light-green; less commonly grayish- to bluish-green or brown.
- Transparent to translucent with a nonmetallic vitreous or, when fibrous, silky luster.
- White streak.
- No prominent cleavage, but breaks by uneven fracture.
- Brittle with a Mohs hardness of 6.5-7.5.
- Usually in long slender to acicular crystals with diamond-shaped cross-sections that are commonly in either parallel or radiated groups; and occasionally in massive fibrous aggregates.
- Fusibility is 7.
- Specific gravity is 3.23-3.27.
- Found in some gneisses and schists.

Staurolite

- Dark-brown, reddish- to yellowish-brown, or brownish-black.
- Translucent to pseudo-opaque with a nonmetallic vitreous to resinous luster.
- White to pale-gray streak.
- No prominent cleavage, but breaks by uneven to sub-conchoidal fracture.
- Brittle with a Mohs hardness of 7-7.5.
- Almost always in short, thick prismatic crystals that commonly occur as interpenetrating cruciform (cross-shaped) twins.
- Weakly to moderately attracted to a neodymium magnet.
- Fusibility is 7.
- Specific gravity is 3.65-3.85.
- Found in some schists.

Talc

- Usually pale-green but commonly also white, greenish-gray, or dark-green (but commonly stained by iron oxides).
- Translucent with a nonmetallic pearly, greasy, or dull luster.
- White streak.
- Prominent cleavage in one direction with easy separation.
- Flexible but not elastic with a Mohs hardness of 1.
- Usually in massive, scaly aggregates; and rarely in pseudo-hexagonal or pseudo-rhombohedral tabular crystals.
- Has a distinctive greasy feel.
- Fusibility is 5-5.5.
- Specific gravity is 2.58-2.83.
- Found in steatite, and some serpentinites and schists.

Tourmaline Group with **dravite**, **elbaite**, and **schorl**

- Most commonly schorl – usually black.
- Occasionally dravite – usually brown, or elbaite – usually shades of green, blue, yellow, pink, or red, and sometimes with multi-color banding perpendicular to crystal length (this is the principal gem variety).
- Transparent to pseudo-opaque with a nonmetallic vitreous to resinous luster.
- Streak of either white or a pale shade of the whole-mineral color.
- No prominent cleavage, but breaks by uneven or conchoidal fracture.
- Brittle with a Mohs hardness of 7-7.5.
- Almost always in long, acicular to slender, prismatic crystals with rounded triangular cross-sections and parallel longitudinal striations on faces, and commonly in radiated groups.
- Schorl is moderately to mainly weakly attracted to a neodymium magnet.
- Fusibility is 3-5.
- Specific gravity is 2.90-3.25.
- Found in some marbles, schists, and skarns.

Vesuvianite

- Usually green, yellow, or especially brown; occasionally blue or red.
- Translucent to pseudo-opaque with a nonmetallic vitreous to resinous luster.
- No prominent cleavage, but breaks by uneven fracture.
- Brittle with a Mohs hardness of 6-7.
- Usually in prismatic crystals with a square cross-section, longitudinal striations, and terminations in the form of a truncated pyramid, and commonly occurring in parallel groups; and occasionally in massive granular aggregates.
- Fusibility is 3.
- Specific gravity is 3.32 to 3.45.
- Found in some marbles and skarns.

Wollastonite

- Usually white to light-gray; less commonly colorless, light-green, yellowish-brown, or brownish-red.
- Translucent with a nonmetallic vitreous to pearly luster or, when fibrous, silky luster.
- White streak.
- Prominent cleavage in one direction, but breaks with a distinctive splintery fracture.
- Brittle with a Mohs hardness of 4.5-5.5.
- Usually in massive granular to mainly fibrous aggregates; and rarely in tabular crystals.
- Commonly fluoresces pale-yellow or orangish-yellow in both short- and long-wavelength UV lights; and commonly exhibits red phosphorescence at both UV wavelengths.
- Fusibility is 4.
- Specific gravity is 2.86 to 3.09.
- Found in some marbles and skarns.

Zoisite

- Usually blue to purplish-blue (called **tanzanite** when gem quality); less commonly colorless, pale-gray, yellow, yellowish-brown, green, or pink.
- Transparent to translucent with a nonmetallic vitreous to pearly luster.
- White streak.
- Prominent cleavage in one direction, but also breaks by uneven or conchoidal fracture.
- Brittle with a Mohs hardness of 6-7.
- Commonly in prismatic crystals; and also commonly in massive, granular aggregates.
- Fusibility is 3-3.5.
- Specific gravity is 3.15-3.36.
- Found in some schists and marbles. Also a rare mineral in some plutonic igneous rocks, although some geologists would consider these to be metaigneous rocks.

Table 5.16: Minerals common in igneous and or sedimentary rocks that also occur in metamorphic rocks.

FELSIC MINERALS[1]
Quartz – found in quartzites, and most gneisses, granulite, and schists.
Potassic and Plagioclase Feldspars – found in most gneisses, granulites, and schists
MAFIC MINERALS[1]
Biotite Mica – found in many gneisses and schists.
Hematite – found in some schists (as the specularite variety).
Hornblende – found in amphibolites, and some gneisses and schists.
Magnetite – found in banded iron formations, and some gneisses and schists.
Muscovite Mica – found in many phylllites, schists, and slates.
Olivine – found in eclogite and some marbles.
Phlogopite Mica – found in some marbles and schists.
Pyrite – nearly ubiquitous, but especially common in hornfels, phyllites, and slates.
OTHER MINERALS[2]
Calcite – found in most marbles.
Dolomite – found in some marbles and serpentinites.

[1] See Table 5.1 for identification tips.
[2] See Table 5.11 for identification tips.

Table 5.17: A simplified classification of metamorphic rocks.

<table>
<tr><th colspan="4">FOLIATED ROCKS</th></tr>
<tr><th colspan="4">SCHIST GROUP (rocks with schistose structure)</th></tr>
<tr><th colspan="2">Texture</th><th>Rock Variety</th><th>Other Characteristics</th></tr>
<tr><td rowspan="3">crystalline</td><td>aphanitic</td><td>SLATE</td><td>• Parent rocks are mudrock or felsic pyroclastic tuff.
• Nearly any color is possible, but usually is bluish-gray to black (from graphite), red to purple (from hematite), green (from chlorite), or brown to yellowish-brown (from goethite or limonite). Can have two or more colors in spotted, mottled, or streaked patterns.
• Micaceous minerals (mainly muscovite) predominate.
• Has a foliation caused by the parallel arrangement of platy clay mineral and mica grains.
• Breaks ('cleaves') easily into thin (<1 cm) flat slabs and sheets with fresh surfaces having a dull to slightly shiny luster.
• Has rare fossils that are usually distorted.
• Can superficially resemble shale but the two rock types are easily distinguished: slate usually cleaves at an angle to any relict stratification (but this is not always present) whereas shale only breaks parallel to stratification; and when struck with a steel tool, slate emits a sharp crack or ringing sound whereas shale only produces a dull thud.
• Mohs hardness is less than 5.5.
• Specific gravity ranges from 2.72 to 2.84 g/cc and averages 2.81 g/cc.
• Gradational with phyllite.</td></tr>
<tr><td>aphanitic to mainly phaneritic</td><td>phyllite</td><td>• Parent rocks are mudrock, felsic pyroclastic tuff, or slate.
• Usually is light-, dark- or greenish-gray to nearly black in color.
• Micaceous minerals (mainly muscovite) predominate.
• Has a foliation caused by the parallel arrangement of platy mica grains.
• Breaks with slight difficulty into thin (<1 cm) flat to wavy or corrugated slabs with freshly cleaved surfaces having a glossy to satin-like or silky sheen.
• Can have porphyroblasts.
• Mohs hardness is usually less than 5.5.
• Gradational with, and transitional between, slate and schist.</td></tr>
<tr><td>phaneritic</td><td>SCHIST</td><td>• Parent rocks are mudrock, felsic pyroclastic tuff, slate, or phyllite.
• Platy minerals predominate, including especially biotite, chlorite, and muscovite micas, and, less commonly, graphite and talc. Usually named after the most abundant mineral: e.g. mica schist. Other metamorphic minerals are usually present.
• Foliation is expressed by a conspicuous, parallel orientation of platy and/or elongated grains. Occasionally there is minor compositional layering as in gneiss.
• Any color is possible but those in shades of gray or green are most common.
• Breaks unevenly along wavy foliation planes, producing irregular flakes and slabs (<1 cm) with freshly cleaved surfaces having a glossy luster.
• Can be porphyroblastic, especially with garnet.
• Mohs hardness is usually less than 5.5.
• Specific gravity ranges from 2.70 to 3.19 g/cc and averages 2.80 g/cc.
• Gradational with gneiss and phyllite.
• Greenschist is the name given to a schist rich in green-colored actinolite, chlorite, and/or epidote. This rock is gradational with greenstone in the non-foliated/granofels group below.</td></tr>
</table>

<table>
<tr><th colspan="4">GNEISS GROUP (rocks with gneissose structure)</th></tr>
<tr><td rowspan="3">crystalline</td><td rowspan="3">phaneritic</td><td>GNEISS</td><td>• Parent rocks are any rock type with abundant silicate minerals, but most commonly granitoid igneous rocks.
• Has conspicuous, parallel, alternating bands or streaks of light-colored felsic minerals (quartz, mainly microcline feldspar, and sometimes sodic plagioclase) and dark-colored mafic minerals (mainly biotite and hornblende). Bands can be straight, wavy, or highly contorted, or replaced by elongated patches or streaks.
• Dark colored bands may exhibit well-developed schistosity.
• Grains are predominantly roughly equant, rather that elongated or platy as in schists.
• Any overall color is possible, but usually is light- to dark-gray or pinkish to reddish.
• Usually does not cleave along the bands.
• Commonly is porphyroblastic.
• Mohs hardness is over 5.5.
• Specific gravity ranges from 2.59 to 3.06 g/cc and averages 2.71 g/cc.
• Gradational with schist and migmatite.</td></tr>
<tr><td>amphibolite</td><td>• Parent rocks are diorite, gabbro, peridotite, pyroxenite, andesite, basalt, or graywacke.
• Compositional banding or streaking varies from poorly to well developed, and consists of light-colored bands of plagioclase feldspar and dark-colored bands of amphibole minerals (mainly actinolite and hornblende). Various metamorphic minerals can also be present, but quartz is always absent.
• Usually dark colored in shades of green, greenish-gray to greenish-black, or brown.
• Mohs hardness is over 5.5.
• Specific gravity ranges from 2.79 to 3.14 g/cc and averages 2.99 g/cc.
• Gradational with amphibole-rich gneiss and amphibole schist.</td></tr>
<tr><td>migmatite</td><td>• Parent rocks are any rock type with abundant silicate minerals.
• Essentially a mixture of gneiss and a plutonic igneous rock, usually with a granitoid composition.
• Has contorted, discontinuous bands of dark-colored mafic minerals (mainly amphibole minerals; the ‘metamorphic’ part) surrounded by lighter-colored patches, veins or discontinuous, non-parallel bands with felsic minerals (mainly quartz and feldspar; the ‘igneous’ part).
• Mohs hardness over 5.5.
• Gradational with gneiss.</td></tr>
<tr><th colspan="4">NON-FOLIATED ROCKS</th></tr>
<tr><th colspan="4">Granofels Group (rocks with NO schistose or gneissose structure)</th></tr>
<tr><th colspan="2">Texture</th><th>Rock Variety</th><th>Other Characteristics</th></tr>
<tr><td colspan="2" rowspan="2">pseudo-clastic</td><td>metasedimentary rocks (non-specific)</td><td>• Parent rock is any sedimentary rock that has been incompletely metamorphosed (i.e. the pre-metamorphic mineralogical, structural, and/or textural features are prominent).
• Attach the ‘meta’ prefix to the sedimentary rock name (e.g. metagraywacke).
• The two most important metasedimentary varieties are treated separately below: metabreccia/metaconglomerate and quartzite.</td></tr>
<tr><td>metabreccia and metaconglomerate</td><td>• Parent rocks are sedimentary breccia or conglomerate.
• Resembles a sedimentary breccia or conglomerate and will not be macroscopically distinguishable from it unless it exhibits an overall greenish cast due to addition of metamorphic chlorite and/or epidote to the matrix. The matrix is essentially the same material as quartzite.
• Mohs hardness is usually over 5.5.</td></tr>
</table>

pseudo-clastic		QUARTZITE or METASANDSTONE	• Parent rock is any grain-supported sandstone, especially quartz sandstone with a quartz-cement, or chert. • Can be of any color, but shades of gray, brown, or red are most common. • Mineralogy is essentially the same as the parent rock, but with added quartz. Although uncommon, metamorphic minerals can also occur. • Usually has a conchoidal fracture with vitreous surfaces. • Usually partially preserves the original clastic texture if from sandstone, but will have a fully crystalline texture if from chert. • Sometimes preserves traces of the original sedimentary stratification. • Breaks through sand grains, rather than around them as in many sandstones. • Although the term 'quartzite' should be reserved only for the metamorphic rock, some geologists and others will apply it to any quartz-cemented sandstone. To avoid any confusion, this rock can be referred to as 'metamorphic quartzite'. • Mohs hardness is over 5.5. • Specific gravity averages about 2.65 g/cc.
		cataclasite	• A catch-all name for any rock exhibiting a cataclastic texture: i.e. consists mainly of angular gravel- or sand-size grains that are broken (crushed or sheared) by brittle deformation. • Associated with fault and shear zones. • Any mineralogy or color is possible, but quartz and feldspars usually predominate. • Distinguished from sedimentary and volcanic breccias by having adjoining angular grains that can be matched because these were originally part of a single larger grain before crushing. • Can be either cohesive or incohesive; consequently, the Mohs hardness varies from less than 1 to over 5.5.
		mylonite	• A catch-all name for any rock exhibiting a mylonitic texture: i.e. consists mainly of smaller angular and larger rounded gravel- or sand-size grains that are ground and distorted by plastic deformation. • Fine-scale layering is commonly present. • Usually cohesive. • Associated with fault and shear zones. • Any mineralogy or color is possible. • Typically finer-grained than cataclasite and can be mistaken for chert or an aphanitic volcanic rock.
crystalline	aphanitic or phaneritic	metaigneous rocks (non-specific)	• Parent rock is any igneous rock that has been incompletely metamorphosed (i.e. the pre-metamorphic mineralogical, structural, and/or textural features are prominent). • Attach the 'meta' prefix to the igneous rock name (e.g. metaandesite or metagabbro). • Gradational with granulite.
	mainly aphanitic	greenstone	• Parent rocks are basalt or any ultramafic igneous rock. • Green to, less commonly, greenish-black in color. • Consists mainly of green minerals: actinolite, chlorite, and/or epidote. If epidote predominates, the rock is called **epidosite**. • Mohs hardness is over 5.5. • Gradational with greenschist in the foliated/schist group above.

crystalline	mainly aphanitic	hornfels	• Parent rock is usually mudrock. • Light- to dark-gray or black in color; less commonly, shades of green, brown, or purple. Commonly exhibits color banding. • It is a hard, dense, nearly homogenous rock, but is occasionally porphyroblastic. • Mohs hardness is over 5.5. When struck by steel, it emits a sharp, ringing sound. • Tends to break by either hackly or conchoidal fracture with freshly broken surfaces exhibiting a dull luster. • Resembles some cherts and felsic to mafic aphanitic igneous rocks, and is difficult to macroscopically distinguish from them.
		lapis lazuli	• Parent rocks are an interbedded sequence of limestone or dolostone, rock salt, and rock anhydrite/gypsum. • Usually consists of lazurite or haüyne (dark-blue or purplish-blue to, less often, greenish-blue) with patches and veins of white calcite and small grains of shiny yellowish pyrite (see Table 3.5 for other mineral attributes and tests). • Mohs hardness for the blue portions is 5-6.
		jade	• Parent rocks are diorite, gabbro, peridotite, pyroxenite, andesite, or basalt. • Predominantly jadeite (a mineral in the pyroxene group) or nephrite (the microcrystalline form of actinolite/tremolite, minerals of the amphibole group). • Usually some shade of green, white, or green with white patches, but other colors are possible (see Table 3.6 for other mineral attributes and tests). • Mohs hardness is 5 to mainly 5.5-7.
		serpentinite	• Parent rocks are peridotite or pyroxenite. • A generally greenish rock consisting mostly of serpentine group minerals with subordinate calcite or dolomite, chlorite, and/or talc. • Mohs hardness is usually less than 5.5. • Specific gravity averages about 2.70 g/cc. • Gradational with steatite and can be confused with jade.
	mainly phaneritic	eclogite	• Parent rocks are basalt or gabbro. • A strikingly beautiful, rare rock that consists mainly of dark-green pyroxene – mainly jadeite (sometimes referred to as omphacite) and/or diopside – and large dodecahedral crystals of red garnet (mainly almandine or pyrope, but occasionally grossular). • Mohs hardness is over 5.5. • Specific gravity ranges from 3.34 to 3.45 g/cc and averages 3.39 g/cc.
		granulite	• A catch-all field term for any non-foliated metamorphic rock that otherwise resembles gneiss in texture and mineralogy (but not structure). • Parent rocks are any felsic or mafic igneous rock. • It can be distinguished from similar-looking igneous phanerites only by the presence of metamorphic minerals. • Specific gravity ranges from 2.63 to 3.10 g/cc and averages 2.85 g/cc. • Gradational with gneiss and phaneritic metaigneous rocks.

<table>
<tr>
<td rowspan="3">crystalline</td>
<td rowspan="3">aphanitic or phaneritic</td>
<td>marble</td>
<td>• Parent rocks are limestone or dolostone.
• Consists predominantly of calcite or, less commonly, dolomite. Test with cold dilute HCl acid to determine which mineral is present or, if both are present, their relative amounts (see section 3.2.8).
• Can be almost any color, but light to medium shades of gray (due to graphite) or white are most common. The color can be uniform or there can be two or more colors in a mottled or vein-like ('marbled') pattern.
• Commonly has mineralogical color banding (commonly grayish from graphite) that can be deformed but this is not foliation but rather relict stratification in the parent rock.
• Is difficult to distinguish from recrystallized limestone or dolostone, or carbonatite, except when metamorphic minerals are conspicuously present. It can also superficially resemble some anorthosites and rock anhydrite/gypsum, but is easily distinguished from them by standard mineral tests.
• Mohs hardness is always less than 5.5 and usually less than 4.
• The above is the correct geological definition of marble, but archaeologists and other non-geologists commonly use it to refer to attractive stones of any mineralogy or origin that take a good polish.
• Gradational with skarn.</td>
</tr>
<tr>
<td>steatite
or
soapstone</td>
<td>• Parent rocks are peridotite or pyroxenite.
• Consists predominantly of talc, which is easily identified by its softness and smooth, greasy feel.
• Common secondary minerals include ankerite, chlorite, dolomite, magnesite, and/or serpentine.
• Usually medium-gray to dark-greenish-gray in color, but when especially rich in talc will be light- to yellowish-gray or nearly white.
• Mohs hardness is always less than 5.5 and usually less than 2. However, if it has been heat treated to increase its hardness, then Mohs hardness will range between 5 and 7.
• Commonly incorrectly referred to by non-geologists as chlorite schist or, simply, chlorite.
• Gradational with talc-schist and serpentinite.</td>
</tr>
<tr>
<td>skarn</td>
<td>• Parent rocks are impure limestone or dolostone.
• Consists almost entirely of calcium-bearing, non-hydrous (i.e. water-poor) silicate minerals of metamorphic origin, including diopside, epidote, garnet (andradite or grossular), tourmaline, tremolite, vesuvianite, and/or wollastonite. Calcite is also commonly present. Usually named after the most prominent mineral: e.g. diopside skarn.
• Mohs hardness can be either less than or over 5.5.
• Called **tactite** when it hosts metallic ore minerals.
• Gradational with marble.</td>
</tr>
<tr>
<td colspan="2">amorphous</td>
<td>anthracite coal</td>
<td>• Parent rocks are lignite or bituminous coal.
• Consists of various hydrocarbon compounds.
• Dark-gray to black in color with a lustrous, sometimes submetallic, sheen and occasionally an iridescent play-of- colors (and then referred to as **peacock coal**).
• Burns with difficulty (i.e. in a blowpipe flame but not in a candle flame), coloring the flame blue and emitting essentially no smoke or odor.
• Tends to break along curved or irregular fractures.
• Handling the rock leaves no dark residue on the fingers.
• Mohs hardness is 2.5-4.
• Specific gravity ranges from 1.3 to 1.8 g/cc.
• Compare with bituminous coal in Table 5.13.
• Converts to graphite at a higher grade of metamorphism.</td>
</tr>
</table>

Table 5.18: Other archaeological 'stones' and tips for their macroscopic identification.

MANUFACTURED MATERIALS	
SMELTED AND REFINED METALS	
gold	• Always native gold either with or without refining. • Rarely pure; commonly has at least a few weight percent silver, unless refined. • Color varies from bright metallic yellow when pure, pale-yellow when 'argentiferous' (rich in silver impurities) to reddish yellow when 'cupiferous' (rich in copper impurities); pale-yellow 'electrum' has similar amounts of both gold and silver. • Ductile, malleable, and sectile. • Shiny yellow streak when pure. • Mohs hardness is 2.5-3 when pure. • Fusibility is 2.5. • Specific gravity is 19.3 when pure; less when impure. • See also Table 3.4 for native gold. □ Used for jewelry and the decorative arts; gilding and sheeting on wood and stone sculptures; tableware; and coinage.
silver	• Either native silver or, more commonly, silver extracted from gold during refining or from galena by smelting. • Color is metallic light-gray to nearly white but, in the presence of water and sulfur, quickly tarnishes to dark-gray, black, or brownish-black. • Ductile, malleable, and sectile. • Shiny white streak when pure. • Mohs hardness is 2.5-3 when pure. • Fusibility is 2. • Specific gravity is 10.5 when pure; can be more or less when impure. • See also Table 3.4 for native silver. □ Used for the same applications as gold except for gilding due to its lower malleability.
copper and copper alloys	• Either native copper or, more commonly, copper smelted from copper-bearing minerals. • Although relatively pure copper was used, this metal was also employed as an alloy with other metallic elements: e.g. with arsenic in 'arsenical bronze', tin in 'tin bronze', zinc in 'brass', silver in 'billion', and others. • Color varies from metallic light-reddish-brown to reddish-orange when pure, pale-gray when rich in arsenic, pale-yellow when rich in zinc, and light-brownish-yellow to yellowish-brown when rich in tin (all darkening with age and exposure); in the presence of water and oxygen corrodes to bluish-green copper oxides (a.k.a. verdigris). • Ductile and malleable. • Shiny red streak. • Mohs hardness is 3-3.5 for pure copper and bronze. • Fusibility is 3. • Specific gravity is 8.94 when pure; can be more or less when impure. • See also Table 3.4 for native copper. □ Used for tools and weapons; jewelry and the decorative arts; coinage; mirrors; tableware; paint pigment; and a glass colorant.
iron and iron alloys	• Either meteoric iron or iron smelted from iron-bearing minerals. • Never pure; always occurs as either a natural or intentional alloy; called 'steel' or 'carburized iron' when alloyed with a few weight percent carbon. • Color varies from metallic gray to black; in the presence of water and oxygen corrodes (rusts) to reddish, brownish, orangish or yellowish iron oxides. • Malleable. • Gray to nearly black streak. • Mohs hardness is 4-8 for modern alloys but probably about 5-6 for ancient alloys. • Fusibility is 6-7. • Specific gravity is 7.30-7.90 when relatively pure, including steel. • See also Table 3.4 for meteoric iron. □ Used for tools and weapons; and many other utilitarian applications.

<table>
<tr><td>lead</td><td>
<ul>
<li>Smelted from lead-bearing minerals, especially galena.</li>
<li>Usually employed as unalloyed lead, although it was rarely pure; also alloyed with tin to produce pewter.</li>
<li>Bluish-gray in color but readily corroding to white lead carbonate.</li>
<li>Gray streak.</li>
<li>Sectile and malleable.</li>
<li>Mohs hardness is 1.5 for pure lead.</li>
<li>Fusibility is ~2.</li>
<li>Specific gravity is 11.4 for pure lead.</li>
</ul>
□ Used for jewelry and small decorative objects; coinage; plates and vessels; weights; sealants and fillings; sheeting, especially for coffins and boat hulls; and a glass colorant when alloyed with antimony.
</td></tr>
<tr><td colspan="2">KILN-FIRED SILICATES</td></tr>
<tr><td>ceramic

[pottery and fired brick]</td><td>
The body form or surface decoration will sometimes easily identify a fragment as coming from a ceramic vessel or brick. The ceramic origin of non-diagnostic fragments can be recognized from the following general characteristics.
<ul>
<li>Composed of heat-induced and -hardened by-products of clay minerals plus other materials added as either a temper or an unintentional contaminant; the most common non-clay materials are quartz silt and sand grains, sand-size rock fragments of many types, volcanic ash, grog (i.e. crushed ceramic wasters from failed firings), crushed mollusc shells, animal dung, and plant chaff (i.e. chopped straw with its impressions commonly preserved in the ceramic).</li>
<li>Exterior colors of most undecorated ceramics are shades of brown or red but occasionally orange or yellow (due to iron-bearing clay and oxidizing conditions within the kiln), but in cross-section the ceramics are commonly color-zoned: brownish or reddish inner and outer surfaces with a dark-gray or black core in between.</li>
<li>A dark-gray or black exterior color, sometimes with a semi-metallic sheen, results from reducing conditions within the kiln, especially if combustible organic matter is added to the clay.</li>
<li>A mixed dark-gray/black and red ceramic is produced by varying the kiln atmosphere from reducing to oxidizing, respectively.</li>
<li>Ceramics rich in added calcite (i.e. marly; from limestone or marble) have a pale-green or pale- to greenish-gray exterior color, and in cross-section are commonly color-zoned with either a dark-gray/black or brownish to reddish core; such ceramics will effervesce when touched by dilute HCl acid.</li>
<li>Ceramics made from iron-poor kaolinite clay are white or pale-gray throughout.</li>
<li>Other colors may be present on the outer, exterior surface due to either the application of a decorative slip prior to firing (i.e. a water-based suspension with a mineral pigment) or painting with a mineral pigment after firing.</li>
<li>Ceramics fired at the highest temperatures become partially or wholly vitrified – i.e. converted to a vitreous, non-porous and impervious, opaque to translucent glass-like material known as 'stoneware' or 'porcelain' – whereas ceramics fired at the lowest temperatures are a non-vitreous, porous and pervious, opaque material known as 'terra-cotta' or 'earthenware'. 'Fired (a.k.a. red) bricks' are another product of the highest firing temperatures.</li>
<li>Terra-cotta/earthenware can be distinguished from stoneware/porcelain by the 'lick test' on a broken edge exposing the interior fabric. Your tongue sticks to the former (because it is porous) but not the latter (as is also true for hard rocks).</li>
<li>Some ceramics have a colored vitreous glaze baked onto their exterior during firing, and these are gradational with faience.</li>
</ul>
□ Used for plates and especially vessels; tiles; fired bricks; oil lamps; pipes; and small sculptures and figurines.
</td></tr>
</table>

glass	• Composed mainly of silica derived from quartz. • Many colors are possible depending on the colorants added during manufacture, but shades of green are especially common in originally colorless glass due to iron impurities. • Isotropic. • Commonly contains trapped air bubbles, especially if blown glass. • Commonly exhibits an iridescent play-of-colors due to reflections off layered scales produced by surface decomposition, but weathered surfaces can also have patches, spots or streaks that are dull, opaque, and usually white. • Commonly exhibits minute surface cracking and pitting. • Can superficially resemble a natural glass like obsidian but, apart from isotropy, obsidian lacks the above properties (see Table 5.8). □ Used for plates and vessels; and jewelry, especially beads, pendants and inlays.
faience	• Composed mainly of silica derived from quartz with added mineral colorants. • Has a white core of semi-fused quartz sand and a colorless or, more usually, brightly colored (commonly blue or green copper based), exterior vitreous glaze. □ Used for jewelry and small decorative objects.
BIOLOGICAL MATERIALS	
PLANT-DERIVED AND COMPOSED OF HYDROCARBONS	
amber	• Composed of a variety of hydrocarbon compounds, principally succinate ($C_6H_{16}O$). • Represents fossilized sap or resin from Conifer trees; similar to frankincense, the unfossilized sap of Boswellia trees. • Mainly brownish, reddish, and yellowish colors, with yellowish-brown and orange most common; rarely blue, green, or black. • Resinous to occasionally waxy luster. • White streak. • Transparent to usually translucent; rarely pseudo-opaque. • Mohs hardness is 2-2.5. • Brittle to sectile. • Conchoidal fracture. • Softens at 150-200°C; a hot point penetrates and produces an acrid, pine-like odor; quickly melts and burns in a blowpipe flame, yielding aromatic fumes; and rubbing with a finger can also usually produce a pine-like odor. • Readily develops a strong frictional electrical charge, and commonly fluoresces yellowish-white (or, less often, blue to grayish-blue) in short-wavelength and bluish-white (or, less often, blue to grayish-blue) in long-wavelength UV lights. • A poor conductor of heat relative to metals and most minerals, and, thus, when taken outside, from indoors at room temperature, amber feels warm in cold weather but cool in hot weather. • May contain entrapped air bubbles, flow swirls, vegetation, insects, etc. • Specific gravity is 1.03-1.10 (floats on a saturated salt solution like seawater but not on fresh water). □ Used for jewelry and small decorative objects.
bitumen [a.k.a. asphalt, tar, and pitch]	• Composed of various hydrocarbon compounds. • Represents solidified residue from petroleum, which ultimately derives from marine (and occasionally lacustrine) plankton, mainly plant (phytoplankton) but also animal (zooplankton). • Black color. • Opaque with a dull or, occasionally, vitreous luster. • Mohs hardness is less than 2. • Quickly melts or burns in a candle flame or under a hot point, giving off dark fumes with a pungent petroliferous odor. • Dissolves in acetone. • Specific gravity is close to 1. □ Used as a sealant for ceramic pots, baskets, and boats; an adhesive for jewelry, tools, and masonry; a varnish for wood, stone, and metal; fuel; medicine; and a preservative for mummies.

<table>
<tr><th colspan="2">ANIMAL-DERIVED</th></tr>
<tr><td colspan="2">composed of hydroxyapatite (a mineral) and collagen (a fibrous protein)</td></tr>
<tr><td>ivory
[a.k.a. tusks]</td><td>
<u>all ivories</u>
<ul>
<li>Composed of collagen with hydroxyapatite ($Ca_5[PO_4]_3[F,Cl,OH]$) plus minor amounts of other compounds.</li>
<li>Represents teeth or dentine.</li>
<li>White to mainly yellowish-white or pale-yellow in color, but developing a brownish-yellow patina with age and exposure.</li>
<li>White streak.</li>
<li>Resinous to waxy luster.</li>
<li>Opaque to rarely translucent.</li>
<li>Mohs hardness is 2-3.</li>
<li>Sectile.</li>
<li>Even to splintery fracture.</li>
<li>Specific gravity is 1.7-2.0.</li>
<li>Non-porous with an outer, enamel-like layer (unlike bone, which it otherwise superficially resembles).</li>
<li>Unaffected by a hot point.</li>
<li>Commonly fluoresces bluish-white in short- and long-wavelength UV lights.</li>
</ul>
□ Used for jewelry and small decorative objects.
<u>Elephant Ivory</u>
<ul>
<li>Tusks have a nearly circular to slightly ovoid outline in cross-section.</li>
<li>Internally exhibits two sets of fine, parallel, crisscrossing lines that are whiter and more opaque than the surrounding material (may require a hand-lens to see) in sections cut at an angle to the longitudinal axis of the tusk; sections cut parallel to this axis will show fine, parallel lines with the same orientation as the axis and sections cut perpendicular to the axis will show concentric rings, especially in the center.</li>
</ul>
<u>Hippopotamus Ivory</u>
<ul>
<li>Tusks have cross-section outlines that vary from nearly circular, roughly square, or triangular with rounded corners.</li>
<li>Internally exhibits one set of fine, parallel, wavy or crinkly lines in all section orientations (may require a hand-lens to see).</li>
</ul>
Although hippopotamus tusks are much smaller than those of adult elephants, they can be similar in size to the tusks of juvenile elephants. There are other kinds of ivory (e.g. boar, narwhale, walrus, sperm whale, etc.). For descriptions of these see Krzyszkowska (1990).
</td></tr>
<tr><td>bone, mammal</td><td>
<ul>
<li>Composed of collagen and hydroxyapatite as in ivory, but with a different internal structure (e.g. has a single large central opening for marrow and many smaller openings for blood vessels).</li>
<li>Mostly from cattle, but also occasionally from deer, goat, horse, and sheep.</li>
<li>Usually white in color, but occasionally yellowish-white and developing a brownish-yellow patina with age and exposure.</li>
<li>Scorched by a hot point.</li>
<li>Internally has a grainy, cellular structure that is best seen on unpolished surfaces; the cells or pores commonly appear as black-filled pits and streaks.</li>
<li>The ‘flat’ bones (ribs and scapulae) have an outer dense cortex and an inner spongy zone; the relative thicknesses can vary greatly and sometimes the inner spongy zone is absent.</li>
<li>The ‘long’ bones (hind and fore limbs) have an outer dense cortex and an inner longitudinal cavity with marrow; the ends are hemispherical or globular.</li>
<li>Short, thick, non-parallel lines may be present internally.</li>
<li>Specific gravity is 1.9-2.1.</li>
</ul>
□ Used for small tools and weapons; and jewelry and small decorative objects, mainly as a non-precious substitute for ivory.
</td></tr>
</table>

<table>
<tr><td>antler</td><td>
<ul>
<li>Similar to bone in its collagen and hydroxyapatite composition.</li>
<li>From deer, elk, and other cervids.</li>
<li>White in color.</li>
<li>Scorched by a hot point.</li>
<li>Has a solid, bone-like outer cortex and a porous, spongy-looking core without any internal laminations.</li>
<li>Has a sub-circular to oval cross-section.</li>
<li>Commonly mistaken for a mammal long bone, but antler lacks the internal graininess and lineation of bone.</li>
<li>Specific gravity is 1.7-2.0.</li>
</ul>
□ Used for the same applications as bone.</td></tr>
<tr><td colspan="2">composed mainly of keratin (a fibrous protein)</td></tr>
<tr><td>horn
&
hoof</td><td>
<ul>
<li>Composed mainly of keratin, the same material as human fingernails.</li>
<li>From antelope, cattle, goat, oxen, and sheep.</li>
<li>Can lose moisture and shrink, resulting in cracking and deformation.</li>
<li>Specific gravity is poorly documented but may be on the order of 1.3-1.4.</li>
</ul>
<u>Horn</u>
<ul>
<li>Usually translucent and nearly colorless in goat and sheep; and translucent and usually either pale-brown, commonly with a greenish tint, or layered and longitudinally streaked with colors ranging from white to reddish brown or black in cattle (buffalo horn is black and opaque).</li>
<li>Has an outer sheath of finely layered material with a cone-in-cone structure, which may be visible on the natural exterior as thin, uneven, circumferential ridges; has a spongy core that is easily removed.</li>
<li>Exhibits a longitudinal, fibrous grain and in transverse (cross) sections is circular to oval in cattle, elliptical in goat, and triangular in sheep.</li>
<li>Varies internally from a solid, dense material with longitudinal tubules at the tip to a central cavity widening toward the base; in transverse sections of the solid portion the tubules appear as oval pores between concentric growth layers with commonly wavy or corrugated lines between the layers.</li>
</ul>
□ Used for drinking vessels; musical horns; containers for gunpowder and other materials; and a wide variety of other utilitarian applications.
<u>Hoof</u>
<ul>
<li>Similar to horn in color, translucency, and internal structure.</li>
<li>Differs from horn in having fine, closely spaced, parallel tubules that vary in cross-section from circular to oval; in longitudinal sections the tubules appear as parallel striations and in transverse sections they look like a dense network of pores.</li>
</ul>
□ Used for glue.</td></tr>
<tr><td>'tortoise' shell</td><td>
<ul>
<li>Composed of keratin.</li>
<li>Shell (carapace) from hawksbill and green sea turtles (not tortoises contrary to the common name for the material).</li>
<li>Has dark-brown to reddish-brown or nearly black mottling or flame-like patterns on a rich- to pale-yellow background.</li>
<li>Translucent.</li>
<li>Specific gravity is 1.25-1.35.</li>
<li>A hot point test yields a 'burnt hair' odor.</li>
</ul>
□ Used for small decorative objects.</td></tr>
</table>

<table>
<tr><th colspan="2">composed mainly of calcium carbonate</th></tr>
<tr><td>red coral</td><td>
<ul>
<li>Composed of magnesium-rich calcite ($CaCO_3$).</li>
<li>Red or pink in color (ordinary coral is white).</li>
<li>Opaque.</li>
<li>Mohs hardness is 3.5-4.</li>
<li>Brittle to sectile.</li>
<li>Even to hackly fracture.</li>
<li>Occurs in branching and bulbous forms.</li>
<li>Non-porous.</li>
<li>Strongly effervescent when touched by cold dilute HCl acid.</li>
<li>Exhibits parallel, longitudinal striations along branches that radiate outward from the center of transverse sections.</li>
<li>Specific gravity is 2.60-2.70.</li>
<li>Commonly fluoresces white in short- and long-wavelength UV lights.</li>
</ul>
□ Used for jewelry and small decorative objects.</td></tr>
<tr><td>pelecypod shell

[a.k.a. bivalve mollusc, or clam shell]

including

mother-of-pearl</td><td>
Pelecypod shells in general
<ul>
<li>Modern and geologically recent shells are composed of aragonite (orthorhombic $CaCO_3$) or rarely calcite (hexagonal $CaCO_3$), but with increasing geologic age, aragonite in fossil shells converts to calcite.</li>
<li>From both freshwater and marine varieties.</li>
<li>Has slightly to highly curved valve fragments, but more complicated shapes come from the hinge area (see also Table 5.14).</li>
<li>Usually is internally laminated with layers of nacre either parallel or inclined to the valve surface, but occasionally exhibits crisscrossing laminations in two directions.</li>
<li>Effervesces strongly in dilute HCl and exhibits other properties of aragonite (see Table 3.6).</li>
</ul>
Mother-of-Pearl
<ul>
<li>Has a strong iridescent play-of-colors that develops on the interior side of oyster shells, a variety of pelecypod (caused by light reflections off overlapping layers of nacre).</li>
</ul>
□ Used for jewelry and decorative inlays.</td></tr>
<tr><td>pearl</td><td>
<ul>
<li>A roughly spherical to ovoid nodule that develops inside the tissue of oysters, a type of pelecypod.</li>
<li>Has a concentrically laminated internal structure.</li>
<li>Composed of aragonite (orthorhombic $CaCO_3$).</li>
<li>Usually white, pale-gray or light-yellow in color, but sometimes pinkish or brownish.</li>
<li>Commonly exhibits a faint iridescent play-of-colors (due to light reflections off layers of platy crystals); can have a 'pearly' luster, but a dull to almost metallic silvery gray luster is common.</li>
<li>Translucent to pseudo-opaque.</li>
<li>Mohs hardness is 2.5-4.5.</li>
<li>Uneven fracture.</li>
<li>Specific gravity is 2.60-2.80.</li>
<li>Effervesces strongly in dilute HCl and exhibits other properties of aragonite (see Table 3.6).</li>
</ul>
□ Used for beads and pendants in jewelry.</td></tr>
<tr><td>gastropod operculum</td><td>
<ul>
<li>Roughly oval-shaped door to the terminal opening of a coiled gastropod shell.</li>
<li>Composed of aragonite (orthorhombic $CaCO_3$).</li>
<li>Exhibits a bold, concentric 'eye-like' color pattern.</li>
</ul>
□ Used for jewelry and amulets.</td></tr>
<tr><td>ostrich eggshell</td><td>
<ul>
<li>Shell from an ostrich egg.</li>
<li>Composed almost entirely of calcite (hexagonal $CaCO_3$).</li>
<li>Consists of three concentric layers collectively less than 3.5 mm thick.</li>
<li>Effervesces strongly in dilute HCl and exhibits other properties of calcite (see Table 3.6).</li>
</ul>
□ Used as a vessel when the eggshell is whole and for jewelry (beads) when the eggshell has been cut into disks.</td></tr>
</table>

Bibliography

Allen, J.R.L. 2017. *Geology for Archaeologists – A Short Introduction*. Oxford: Archaeopress.

Andrefsky Jr., W. 2005. *Lithics – Macroscopic Approaches to Analysis* (2nd. edn). Cambridge: Cambridge University Press.

Berry, L.G., B. Mason and R.V. Dietrich 1983. *Mineralogy: Concepts, Descriptions, Determinations* (2nd edn). San Francisco: W.H. Freeman and Co.

Brush, G.J. 1905 (revised by S.L. Penfield). *Manual of Determinative Mineralogy with an Introduction on Blowpipe Analysis* (16th edn). New York: John Wiley and Sons.

Clark Jr., S.P. 1966. *Handbook of Physical Constants*. Memoir 97. New York: The Geological Society of America.

Clough, T.H.McK. and A.R. Woolley 1985. Petrography and stone implements. *World Archaeology* 17/1: 90-100.

Deer, W.A., R.A. Howie and J. Zussman 2013. *An Introduction to the Rock-Forming Minerals* (3rd edn). London: The Mineralogy Society.

Dietrich, R.V. and B.J. Skinner 1979. *Rocks and Rock Minerals*. New York: John Wiley and Sons.

Fettes, D. and J. Desmons (eds) 2007. *Metamorphic Rocks: A Classification and Glossary of Terms. Recommendations of the International Union of Geological Sciences*. Cambridge: Cambridge University Press.

Gleason, S. 1960. *Ultraviolet Guide to Minerals: A Complete Working Manual for the Use of Ultraviolet Light in Locating and Recognizing Minerals, Including Field Identification Charts*. Princeton: Van Nostrand.

Gribble, C.D. 1988. *Rutley's Elements of Mineralogy* (27th edn). London: Unwin Hyman.

Harrell, J.A. 2024. *Archaeology and Geology of Ancient Egyptian Stones* (2 vols). Oxford: Archaeopress.

Herz, N. and E.G. Garrison 1998. *Geological Methods for Archaeology*. New York/Oxford: Oxford University Press.

Hodges, H. 1964. *Artifacts: An Introduction to Early Materials and Technology*. London: John Baker.

Hurlbut Jr., C.S. 1966. *Dana's Manual of Mineralogy* (17th edn). New York: John Wiley and Sons.

Kempe, D.R.C. and A.P. Harvey (eds) 1983. *The Petrology of Archaeological Artefacts*. Oxford: Clarendon Press.

Klein, C. and B. Dutrow 2007. *Manual of Mineral Science (after James D. Dana)* (24th edn). Hoboken: John Wiley and Sons.

Krzyszkowska, O. 1990. *Ivory and Related Materials – An Illustrated Guide*. Classical Handbook 3, Bulletin Supplement 59. London: Institute of Classical Studies.

Le Maitre, R.W. (ed.) 1989. *A Classification of Igneous Rocks and Glossary of Terms – Recommendations of the International Union of Geological Sciences, Subcommission on the Systematics of Igneous Rocks*. Oxford: Blackwell.

Luedtke, B.E. 1992. *An Archaeologist's Guide to Chert and Flint*. Los Angeles: Institute of Archaeology, University of California.

Matthews, W.H. 1960. *Texas Fossils – an Amateur Collector's Handbook*. Guidebook 2. Austin: Bureau of Economic Geology.

Moses, A.J. and C.L. Parsons 1916. *Elements of Mineralogy, Crystallography and Blowpipe Analysis* (5th edn). New York: D. Van Nostrand Co.

Mottana, A., R. Crespi and G. Liborio 1978. *Simon and Schuster's Guide to Rocks and Minerals*. Milan: Arnoldo Mondadori Editore.

Odell, G.H. 2004. *Lithic Analysis*. New York : Kluwer Academic/Plenum Publishers

Ogden, J. 1982. *Jewellery of the Ancient World from the Beginning of the Bronze Age to the End of the Roman Period*. New York: Rizzoli.

Pellant, C. 1992. *Eyewitness Handbooks – Rocks and Minerals*. New York: Dorling Kindersley.

Photos-Jones, E. and A. J. Hall. 2011. *Lemian Earth and the Earths of the Aegean – An Archaeological Guide to Medicines, Pigments and Washing Powders*. Glasgow: Potingair.

Rapp, G.R. 2002. *Archaeomineralogy*. Berlin/New York: Springer.

Raymond, L.A. 1984. *Petrography Laboratory Manual. Vol. 1 – Handspecimen Petrography*. Boone: Geology Services International.

Rosenfeld, A. 1965. *The Inorganic Raw Materials of Antiquity*. New York/Washington: Frederick A. Praeger.

Shackley, M. 1977. *Rocks and Man*. New York: St. Martin's Press.

Siddall, R. 2018. Mineral pigments in archaeology: their analysis and range of available materials. *Minerals*, v. 8, n. 5, article 201. https://doi.org/10.3390/min8050201

Sinkankas, J. 1972. *Gemstone and Mineral Data Book*. New York: Winchester Press.

Spock, L.E. 1962. *Guide to the Study of Rocks* (2nd edn). New York: Harper and Row.

Streckeisen, A.L. 1967. Classification and nomenclature of igneous rocks. *Neues Jahrbuch fiuur Mineralogie Abhandlungen* 107: 144-240.

General index

Mineral index

Rock index